Dr. Sanders' Guide to
Boosting Dairy Profits

Donald E. Sanders, DVM

Editor: Paul W. Pratt, VMD
Associate Editor: Ann B. Bennett
Consulting Editor: Meg Gaige
Production Manager and Cover Design:
Elisabeth S. Stein

American Veterinary Publications, Inc.
5782 Thornwood Drive
Goleta, CA 93117

Cover Photo: Larry Lefever, Grant Heilman Photography

Library of Congress Catalog Card Number 89-84351
ISBN 0-939674-26-2

Printed in the United States of America

Author's Note

The seed for this book was germinated in the early '80s when many of my clients were despondent about the bleakness of the agricultural economic picture. Despite the doldrums, I was determined to show my clients they could make a profit if they really concentrated on improving production and cutting costs. Many times, they just cut costs without looking at what the potential payback might be.

I got on my soapbox and began telling my clients that they needed to look at their dairies closely and improve their performance while keeping costs in line, rather than just cutting back to bare-bone necessities. These writings are a summary of the ways I saw my clients improve their operations and, in addition, a retrospective look at some of our mistakes. I learned from these mistakes and I think you will also.

This book would not be possible without what I refer to as "the women in my life." They include my wife, Judy, who faithfully works as hard as any wife, mother, veterinarian, friend and partner could. My daughter, Michelle, who, at the age of 15, is as good as an administrative assistant with a string tied on my finger to keep me organized with the list of what I am to do next.

Our very capable office staff includes Karen, who has been the drill sergeant that has kept our office straight, and Ginger, always dependable and always interested, deserves special mention. Bea spent 18 years with us and, after over a year's retirement, has returned to continue as a staunch ally, friend and employee. A shiny gold star goes to Eugenia Snyder, the person that got this project up and running. Eugenia helped plan the book and develop the outline into a 100-page guide, rather than the topics that I had listed on the back of a couple of used envelopes. Her input started from my treatment of her horse 3 times over New Year's Eve and the following day.

Meg Gaige is the editor who continued the fine job that Eugenia started, rapping me on the knuckles for bad grammar when it was required like a school marm with a ruler. A prequisite I specified before starting was that she couldn't rap my fingers on my palpating arm. Never once did she hit me and she was always quick to suggest a new idea or a different slant to my musing.

My sons, Allen and David, ages 10 and 12, respectively, also patiently waited for me to get over this obsession of writing a book. I am indebted to Dr. Paul Pratt and his staff at American Veterinary Publications, who took a chance on an old, ornery, procrastinating, veterinary rascal with unknown writing ability. They made a silk purse out of a sow's ear. Other people too numerous to list encouraged me or added to my determination to finish this project. Though unnamed, they know who they are and have my heartfelt thanks.

Finally, my resolve was strengthened by a very special walk I made to Emmaus that gave me the insight and dedication to do this project. Without the establishment of those God-given ideals and priorities, I wouldn't have ever finished.

Don Sanders, DVM

About the Author

Don Sanders developed a keen appreciation for dairying and an interest in veterinary medicine when he was a boy on his family's dairy farm in northwestern Ohio. He attended Ohio Northern University as an undergraduate and received his doctorate from Ohio State University's College of Veterinary Medicine in 1968. Following additional field training and written examinations, Dr. Sanders became a Diplomate of the American College of Theriogenology.

Dr. Sanders and his wife, Dr. Judy Sanders, established the Urbana Veterinary Clinic in Urbana, Ohio, in 1968. In 1974, they built an award-winning veterinary hospital that is home to the clinic today.

In addition to maintaining a busy large-animal practice, Dr. Sanders has conducted field research on BVD and mycoplasma vaccines, stray voltage, mastitis, toxic substances and production medicine. He is consulted by dairymen and veterinary practitioners from across the US.

Dr. Sanders' commitment to his clients, patients and family is untiring. His professionalism is tempered with warmth and wit – and a devilish desire to play a practical joke now and then. To those whom he serves, he is a genuine friend and a bright ray of sunshine on a gloomy day.

Contents

Foreword

U.S. dairy cattle are in demand because of their ability to efficiently produce milk under a variety of climatic conditions. They are the largest contributors of superior dairy genetics and hold the world's records for milk, butterfat and protein yield. Whether a breeder is developing a herd to produce more milk, butterfat and protein, or cows with superior functional type, the U.S. offers the best selection of dairy cattle in the world.

Every year more of the world's dairy producers COME TO THE SOURCE – and import dairy cattle, semen and embryos from the United States. The U.S. Purebred Dairy Cattle Association (PDCA) provides technical information and assistance to help new buyers succeed with U.S. dairy genetics. The PDCA is a strong proponent of comprehensive herd health programs and sound management practices; unless cows are managed properly by well-trained personnel, they will not reach their genetic potential. Lack of knowledge and experience impedes adaptation of improved dairy genetics and evolution of successful, modern dairy enterprises.

The PDCA is delighted to offer *Dr. Sanders' Guide to Boosting Dairy Profits* to dairy farmers worldwide. This book outlines a practical approach to dairy herd management and will help dairymen develop management practices that can translate into increased milk production and profits.

The author, Dr. Donald E. Sanders, is a graduate of Ohio State University's College of Veterinary Medicine and is a Diplomate of the American College of Theriogenology. Dr. Sanders has been in private practice since 1968. Because of his expertise in the areas of dairy cow nutrition, reproduction, disease processes and production medicine, he is consulted by clients and veterinary practitioners from all over the U.S. He has authored more than 50 articles for professional and farm industry publications and is widely recognized as an expert on dairy cattle. Dr. Sanders and his wife, Dr. Judy Sanders, have built a successful large and small animal practice at the Urbana Veterinary Clinic, an award-winning facility they designed and built in Urbana, Ohio.

Dr. Paul Pratt and his staff at American Veterinary Publications have transformed Dr. Sanders' text into a handy, easy-to-understand guide. The book contains more than 100 photographs and illustrations that effectively complement the text, including line drawings by Joe Sample and cartoons by Dr. Robert M. Miller.

Dr. Sanders' Guide to Boosting Dairy Profits stresses the importance of developing a herd health program and provides invaluable information on every aspect of dairying, from keeping records and setting reproduction and production goals, to feeding, milking and managing calving and a range of diseases. It describes management practices necessary to ensure that cows with superior genetics meet their production potential and, at every turn, points the way to increased profits from dairy farming.

Donald V. Seipt, President
PDCA

PDCA

The U.S. Purebred Dairy Cattle Association provides leadership within the dairy industry on issues of importance to breeders of purebred dairy cattle. The Association develops programs to increase the acceptance of purebred dairy cattle, while expanding opportunities for herd owners to profit through improved management and marketing. PDCA is comprised of the 6 major dairy breed organizations in the United States:

American Guernsey Association
American Jersey Cattle Club
American Milking Shorthorn Society
Ayrshire Breeders' Association
Brown Swiss Cattle Breeders' Association
Holstein Association of America

PDCA International Market Development Council

The PDCA International Market Development Council represents all facets of U.S. dairy genetics including the exporters of those genetics. The Council conducts market development activities with the U.S. Foreign Agricultural Service promoting U.S. dairy genetics worldwide. Council members include:

American Embryo Transfer Association
American Guernsey Association
American Jersey Cattle Club
American Milking Shorthorn Society
Ayrshire Breeders' Association
Brown Swiss Cattle Breeders' Association
Holstein Association of America
Livestock Exporters' Association
National Association of Animal Breeders
State Departments of Agriculture

COME TO THE SOURCE

U.S. Purebred Dairy Cattle Association
PO Box 208
Beloit, Wisconsin 53511-0208 USA

1

Not Enough Magic Bullets

This book's purpose is to show dairymen how to make cows healthier and more profitable. My veterinary practice, in Ohio and surrounding states, includes clients who are the best people in the world. But they have some quirks.

They always want to know what the new "magic bullet" is. The magic bullet is the shot that will get all cows to come into heat and get pregnant, the shot that will stop pneumonia, the shot that will make cows immediately healthy.

But the new magic bullets aren't what you would expect. They're different from the antibiotics we've often considered magic during the past 30 years. Cows easily get pregnant with these new bullets. Milk production can increase with their liberal use. In short, they not only make cows healthier, but also make your finances healthier.

You can't buy them; they aren't bottled. They're called production medicine or herd health, "attention to the details of good management," "going the extra mile," and assessing management decisions in terms of economic return. Let me start with a true story.

An Expensive Oversight

One of my young clients, Michael (not his real name), was concerned about higher production. His cows had a 16,000-lb rolling herd average. He spoke wistfully about Jerry, another of my clients, whose herd had a 22,000-lb rolling herd average.

Michael mentioned that Jerry had Harvestore silos. He also had a total mixed ration and a computerized feeder. Michael believed if he had a couple of Harvestores and a computer, his cows could milk like Jerry's.

"Let's look a little deeper," I told Michael. Tactfully, I tried to compare Michael's 16,000-lb herd with Jerry's 22,000-lb herd. Michael was content to feed his cows a complete feed from a local company. He didn't do feed analyses.

Jerry, on the other hand, balanced his ration for soluble and insoluble protein. He also used several protein sources, depending on current costs. Jerry used a nutritionist and worried about optimum mineral levels. He not only looked at calcium and phosphorus, but also was concerned with trace minerals.

In addition, Michael's cows were palpated only when he had time. He was fairly regular in calling me out to do this, but he didn't call during corn planting or harvest. At one point, I hadn't seen his cows in 2 months.

Jerry had a standing appointment for reproductive work. I palpated every 3 weeks. He kept records on a breeding wheel and computer printout from the Dairy Herd Improvement Association (DHIA). He was in the process of developing a method to have his own computer terminal access a mainframe computer at a university.

Michael, in contrast, kept his own records on the back of paper towels attached to a clipboard. I also mentioned to Michael that he tended to call me only after his cows had been sick for a couple of days. I saw Jerry's the same day they first seemed "off."

Michael understood what I was telling him, and I think he felt depressed, but I pointed out that he was a better manager than when we first started working together 5 years ago.

He no longer had a high somatic cell count (SCC). His calf mortality was less than 1%, as compared with earlier losses of more than 35%. I told him the important thing was that he set goals and work toward them. If he didn't strive to raise production and keep costs in line, he might not survive economically.

My clients will tell you that I cajole them, coach and nag them, and regularly try to rattle their chains. But they make money from my health recommendations. If they don't, I know I'll lose them as clients.

When I visit cow herds across the United States, I see that profitable, high-producing herds have one common denominator: healthy cows. When you think of a cow, think of a sawhorse. Put enough weight or stress on it and it will eventually break. Put nutritional stress on the cow, add calving stress, stray voltage, overmilking, bad ventilation, poor housing and extreme weather conditions, and she "breaks" too. She gets mastitis, doesn't breed, milks poorly or gets sick, none of which make her profitable.

In essence, I'm talking about a production medicine program for your cows — a planned maintenance program to keep your cows milking for the most net profit (see Chapter 4). That's the difference between Michael's and Jerry's cows.

Your vet's regular visit should include more than palpation. It should include procedures that help prevent disease and improve performance. Review milking procedures to prevent mastitis, but more important, cut the SCC so that production goes up. Invite a milking equipment specialist to evaluate the milkers at least twice a year. This may be someone other than the installer.

Check your calf-rearing program for housing and nutritional faults. Test cow rations and bring them into balance for optimum net return. Some dairymen hire an independent nutritionist for this task.

Your vet can help you select the right experts for your needs. Sometimes, he or she can fill the roles of milking equipment expert and nutritionist. If you decide on outside consultants, plan on a team effort.

Meet as a group at least twice a year, and do what you can to enhance communication among members of *your* team. You might need to meet more frequently if things aren't going well.

Once you're ready for these steps, make an appointment when your vet is not flooded with calls. Tell him or her you want to improve your herd's health program. You want it to be of larger scope than just health; you want it to be a production medicine program. If your vet doesn't seem interested in a planned health program, find one who is.

I originally wrote the story of Michael and Jerry for *Farm Journal Dairy Extra.* Now I want to share with you the "rest of the story," as Paul Harvey would say.

> *Michael's ascent to profitability was filled with road blocks. He continued to rely on a complete feed from a feed company. He arranged for some feed tests, but his complete feed wasn't very adaptable. So he wasn't able to implement much of the trace mineral program that was necessary to correct his breeding problems. His calving interval hovered around 14 months.*
>
> *Michael had never checked his milking equipment to ensure it was doing a top-notch job. Without asking anyone independently about some new milker equipment, he purchased an adaptation for his equipment because, "the salesman knew what he was talking about." Michael got his Harvestores and the computerized feeder. He was proud as punch because now he was going to be able to do the tremendous job of management required for 20,000 lb of milk.*

Michael and the salesman had figured out his cash flow in feed savings that would result from the high-quality feed coming out of the blue silos. I couldn't find fault with those silos and the potential for top-quality haylage. What Michael never realized was that to get top-quality feed out, you must put top-quality feed into those silos. He wasn't able to do this to the degree required to cut his protein costs and make the cash flow.

Michael figured that he would need 20 extra cows for his cash flow. He calculated feed storage large enough to accommodate the extra cows so he wouldn't run out of feed. What he hadn't figured was the stress created by adding cows to his already full barn. The new heifers had to stand in the mud behind the barn. He had no provisions for feeding them other than to carry feed to them.

Michael was lacking roomy dry-cow facilities and had to use what space he had for the extra milk cows he added. His dry cows were turned out with the heifers behind the barn. He then had the dilemma of correctly feeding the growing bred heifers or the dry cows. The result was the dry cows got thin and the heifers were stunted because they were crowded out by the dry cows. Michael was short of bunk space for both.

Because the fresh cows were thin and lacking a quality dry-cow program before freshening, they milked poorly at freshening. Cash flow got tighter, so Michael let his hired man go. This meant even more work for him. Michael's wife came to the barn and worked right beside her husband night and day. With her help, he was barely able to keep up with the chores. When spring came, none of the crops got in on time. Corn planting ran into haylage harvesting, and the result was poor-quality haylage. Milk production dropped further. Michael and his wife worked harder to get their operation back on track. I tried to be supportive but I couldn't feed and milk and do their everyday work for them. My support consisted mainly of trying to get sick fresh cows going and treating heifers stunted by pneumonia.

Michael had me consult with nutritionists to balance his ration. But recommendations weren't ever used because Michael couldn't get far enough ahead to purchase the feed necessary. He used his computerized feeder, but ran out of feed often enough that the cows used it poorly.

Finally, Michael filed a bankruptcy petition in federal court. He has lost the farm he inherited from his parents. His *wife's health is precarious. Michael now has a construction job that pays $8.50 an hour.*

Several things can be learned from this story. There was no doubt about Michael's and his wife's dedication to working hard on their farm. But he wasn't patient enough to develop finesse in his management before moving on to bigger things. He originally started by purchasing a 20,000-lb average herd from his neighbor. When he kept the program simple and only milked 30 cows, he made money. The straw that broke the camel's back was trying to expand before he mastered the basics.

Jerry, meanwhile, has gained another 1500 lb of production on his rolling herd average. His SCC is less than 100,000 and his calving interval is just under 13 months. But his success hasn't been without trials. Some moldy brewer's grain delivered to his farm really threw his cows off. He had 2 abortions and several fresh cows that severely scoured. Yet Jerry had a team working for him. While he was very concerned, his consultants helped him work through the problems efficiently. Today, his cows are averaging 77 lb/cow. At last test date, he was milking 70 cows.

Today when I drive past Michael's old farm I see 2 blue monuments still standing as sad reminders of the false American dream that we can merely spend our way to good management.

What follows is intended to replace that illusion with practical facts. This information, if applied, can lead to a healthy, well-managed and profitable herd of dairy cattle.

2

Setting Realistic Goals

John's Folly

John Jones was in New York City. He wanted to go to Boston, so he went to the station and bought a ticket. Having a few minutes to spare, he walked over to some scales, stepped on them, inserted a coin, and out came his fortune: "Your name is John Jones, you weigh 188 lb, and you are going to catch the 2:20 to Boston." Jones was certainly puzzled. Sensing a trick, he decided to "fool" whoever or whatever was responsible. He went into the men's room, changed his clothes and put on sunglasses as a disguise. Once again he stepped on the scales, inserted his coin, and out came his fortune: "Your name is John Jones, you still weigh 188 lb, but you just missed the 2:20 to Boston."

This book is for those of you who have missed getting your cows on the 2:20 to Boston. Or maybe they got off track before they made it to their destination. I'm convinced that you can make your cows milk for top profits. But to do so, you have to lay the groundwork and follow the steps outlined in this book. You can't take shortcuts. A baseball player is out if he doesn't touch all the bases; you'll be "out" if you try to skip any of the steps to having a top herd.

The Steps to Success

The first step is to feel good about being a dairyman. You like where you are and you don't feel trapped with your cows. The second step is to recognize the value of the people around you. Your spouse or parents can be supportive even if they aren't doing the physical labor. Your children love you even though you feel like you don't spend enough time with them. It's a necessity to work and live with them; be supportive of them and it will return to you.

The third and crucial step in achieving maximum profits and top production is to have a plan or a set of goals. You need a blueprint to build a house. It's even more important to have a blueprint for your herd. Goals should spell out in specific steps how you plan to get to the top of the heap.

I'll bet you're going to argue, "Well I'm doing the best I can with what I have." *Are* you? Take a bar of iron and use it for a door stop and it's worth about a dollar. Take that bar of iron and make horseshoes and it's probably worth about $50. Take that same bar of iron and remove the impurities, refine it into steel, then manufacture it into watch springs for precision watches. It'll be worth a quarter of a million bucks.

How about your cows? Over the block, they're worth salvage prices. Freshen them out and feed them according to NRC requirements, and they'll be worth heifer replacement prices. Or feed them a well- balanced ration, bypass protein and protected fat, get them bred on time and put them on DHIA test. Now you've got something!

For starters, let's divide your dairy business into the fundamental building blocks that go into improving net profits.

Keeping Score

Is your incidence of milk fever less than 3%? Are cases of ketosis and displaced abomasum each less than 3%? Are fewer than 5% of your calves born dead? Is your incidence of retained placenta also less than 5%? If you can answer yes to these questions, chances are your cows are milking at their best. If you don't know, how are you going to keep score? Let's start a tally so you know the score.

Are all of your cows in heat by 45 days after freshening? Most of them will have had to cycle by then if you hope to breed them the first time by 60 days. It might sound lofty, but I shoot for servicing the first time by 60 days. You can expect these cows to cost you $2 per day for every day they are open past 100 days. If you have to breed them 2 or 3 times, it's easy to see they'll pass that 100-day marker. Service them early if you also want to achieve a 12.5-month calving interval.

All of these reproductive performance goals trace back to seeing your cows in heat the first time by 45 days after freshening. If you don't get that right, there isn't much chance of reaching the rest of the goals. So, if you aren't marking down heats on fresh cows and keeping tabs on them for the next heat, I suggest you sneak out to the breeding records and set up a method of doing it before somebody asks you why you're not!

The next category to include in your goals for improved breeding performance is services per conception. An AI average across the country is 2.0 services per conception. Try to do better.

Today many dairy processing plants pay a premium for milk with a SCC of less than 300,000. This is a very practical goal. If you achieve it, the reward immediately shows up in your paycheck. What a "high" that can be. Try to keep clinical cases of mastitis less than 10% on a herd basis. Do you have a tally so you know the score on mastitis cases?

Aim for an involuntary culling figure of less than 15%. I define involuntary culling as having to sell a good cow because you didn't have the choice of keeping her. This is the cow that has mastitis or won't catch. Conversely, do enough voluntary culling so you move nonperforming animals out of the herd. In most herds this number should be 10% or more of the herd per year. Too often I hear clients say they haven't culled many animals because they've had a string of bull calves. Usually I reply with a question: "How many heifers died before they had a chance to freshen?"

Next, let's talk about calves that hit the ground alive. Do you lose more than 1% of the crop in the first 30 days? And more than 2% between day 30 and freshening? A quick look at calf mortality records will give you the answer. Care for heifers so that they calve at 24 months of age and weigh 1,200 lb at that time (in large breeds). To achieve that goal, you'll need to look at your heifers more often and use a weight tape on them.

How about some goals for your ration? Can you produce milk for less than $5/cwt? If you're like my clients, I'll bet you've never calculated it other than with some feed salesman or the extension service giving book values.

Set Your Goals

There are many other goals you can make for yourself, such as the cost of raising heifers, breeding all of your heifers AI, or maintaining optimum body condition score in the dry cows. Whatever your goals, make sure you write them down! Post them on the switch where you turn the milker on, or maybe the mirror you look at when you drag yourself out of bed. If you're really brave, show them to your vet or your herdsman. The positive aspect is that you will get support as your herd improves. The negative aspect is that you put pressure on yourself that can cause public humiliation. You may want to keep your goals private if they could become a source of embarrassment. However, if they are a potential source of embarrassment, maybe you're initially setting unrealistic goals. Start with smaller ones and win some early victories before you go for the bigger ones.

While you feel the urge to "go for it," set some guidelines. Write down where you're going to start. If this book gets you thinking, set a goal of reading so many pages a week. Be realistic. Don't try to rush through; let the ideas you read simmer. Make notes to yourself on ideas that might fit your operation and you will be off and running!

3

Keeping Records

There is probably no topic less exciting than keeping records. I think some of my clients would rather clean the barn with a fork than keep their records current. Yet, those who are making money have a firm grasp of their costs as well as the performance of their cows. Records will save the day if you use them in all facets of management.

Why Keep Records?

You can enhance the reproductive performance of your cows with good records. Records improve heat detection by allowing you to anticipate open cows' heat cycles, and then reduce the interval to first heat. As a result, the interval to first service also improves so you can reduce the number of average days open. This often reduces services per conception. You will have lower semen costs and expend less labor getting your cows pregnant. The net result is that you may gain a lactation per cow over several seasons of reducing the average days open.

Good records help you identify problem breeders before they get so far into their lactation that it's uneconomical to redeem them. This reduces services per conception and average days open. The result is lower culling for those genetically superior cows that are always hard to cull.

Records show when cows should be dried off. It's critical that cows have a normal dry period. If it's too short, they aren't in proper flesh to milk hard the next lactation. If it's too long, they get fat and develop the complications of fat cow syndrome.

Good records allow you to anticipate when cows will freshen. Then you can get them on a prefreshening ration 2 weeks before calving.

Detailed records provide critical data for your vet about individual cases or a herd problem. This improves the speed and accuracy of diagnosis. Otherwise, your vet may need several trips to diagnose a tough problem. With detailed records, your vet can speed treatment and recovery of the cow, and improve your herd's performance. This can reduce production losses and often reproduction losses, as well as reduce vet bills.

Involuntary culling rates invariably go down, so you have more options regarding culling. If you are like me, you feel better about selling a cow that is milking poorly than a good cow you can't get bred.

The bottom line is improved herd health and increased production. The increased production is almost all profit because there is very little increased cost associated with good records.

Find the Problem Areas

Records provide a good foundation for pinpointing areas in which you need to improve. They provide the data to identify potentially weak areas in your breeding program. Looking at basic categories, such as average days open, often provides insight. If your average days open is high, then you know to look at the heat detection program and evaluate whether enough time is being spent in heat detection. SCC records are a good place to look for cows with chronic mastitis that should be culled. Records point out the inefficient breeding cows, and the chronically sick ones that should also be culled.

Records are essential to measure your herd's production. Rolling herd averages indicate in which direction production is headed. Records point out cows that are good protein and butterfat testers. Your breeding program can be channeled toward breeding for protein and butterfat. Because both of these characteristics are highly heritable (passed on to offspring), records direct your efforts toward optimizing them. In short, records can show if individual cows, the herd and you are doing a good or a not-so-good job. And they are a way of acknowledging a good job done by employees in keeping the herd on track.

Keeping production information with DHIA or DHIR is likely to stimulate breeding stock sales. This information is also used to determine production costs and profit threshold. It's also important for management/investment decisions. Often, some of these figures are necessary for a financial institution's support. A loan application always has more validity if there are records confirming the need for the loan. In addition, the data may be necessary to verify tax consequences of your income.

It's important to have detailed records, but they are only as good as their accessibility. If you keep them at your house or at your accountant's, you may not use them often. While it's important to have detailed records at one of those locations, you really need a day-to-day summary at your fingertips. A mini-office or cabinet is a must at the dairy barn. A ledger or diary must be available for daily entry and review.

Individual Health Records and Cow History

I don't know of a single dairyman who can effectively manage his herd without individual cow records. I have some clients who try, but their management skills are severely handicapped.

You should have at your fingertips all pertinent information about each cow: her birth date, freshening dates and breeding dates. Many of my clients don't keep heat date records. If they did, heat detection would be much more effective. They could anticipate heats and achieve a smaller number of days open. Drying-off dates are necessary to plan for dry periods. More important, drying-off dates are critical for proper dry cow therapy. You must closely watch the dry treatment interval so a problem cow doesn't pop up with antibiotic residues in her milk or her meat if you change your mind and decide to cull her.

Every herd should have a vaccination program. The dates of vaccinations must be recorded to ensure that cows are boostered at the proper time for optimum immunity. Heifers need regular deworming. Later, I will discuss the tremendous return you get from proper deworming of replacement heifers (see Chapter 8). For this profitable management practice to work, you must know the age and status of your heifers.

All illnesses should be recorded for future reference by your vet. Culling decisions may be based on previous history of illness. Included with the record of illness is such information as signs, abnormal behavior, drug reactions, diagnosis and treatment. Be sure your vet writes down his findings as legibly as possible.

Various record systems are available. I have clients who use several different systems. They may have a notebook system or individual cow folders. They also ask me about the advantage of a breeding wheel or barn charts. Then they vaguely mention that the extension office has been encouraging them to use a computerized record service. The question I am asked is, "Which one should I pick?"

I usually make it more difficult because I tell them they should use all of the above. The cow folders or 3-ring binder can be used for individual cow records. A small extra notebook can be stuffed in a back pocket and carried. Cow folders can be kept in an office close to the treatment area for easy access, and updated from the notebook.

Every dairy I work with has a breeding wheel or needs one. It should be easily accessible for following up on the anticipated heat dates, pregnancy check time, drying-off date, and freshening date for each cow.

Your cows should also be on a testing program, whether it is DHIA or owner-sampler records. I don't think you can do without a production measurement system. This is critical if you are to have goals for your cows. With the production records, most record centers also provide data on average days open, calving interval and conception rates.

Very large herds may find a breeding wheel awkward and cumbersome for records on each cow. I recommend that they have a computerized record system, with a barn terminal connected to a mainframe computer.

Herd Summaries

To have a firm grasp on the management of your cows, you need information that has been summarized for all of them. This is the only way to know the results of your efforts. Your system can be homemade or developed expressly for dairy herds. Homemade systems can be relatively simple, and may consist of pins on wires, a pegboard system or a blackboard.

In my opinion, every dairyman should be looking toward computerizing his records, either through the extension service or his own equipment. If you are to survive and prosper in the coming years, your information must be current. I recommend that you view any homemade system as a temporary measure until you can computerize your records.

Regardless of your record-keeping system, the information must be accessible when you need it. You should have it available when your vet visits. Make as efficient use of his or her time as you can and you will get more in return. If possible, provide your vet with a printout of your records when they arrive back from the DHIA information center. It will make him or her feel more like part of the team. Sometimes, your vet can offer management observations from your records that are very beneficial. You will find that you and every other member of your production improvement team will benefit from keeping better records.

4

Developing a Herd Health Program

In the past, a veterinarian and dairyman had an arrangement: if a cow became sick, the veterinarian was called to examine her. The doctor was trained in physical examination; long on formulas or concoctions and short on effective antibiotics to treat her. Often, the vet was a family friend and available at all hours of the day and night to deliver calves, treat colic and minister to milk fever cases.

In the '60s and '70s, this relationship began to change. Dairymen began to deliver their own calves and treat their own milk fever cases. Vets realized that if herds were to be profitable, more was needed than drug therapy. Many started scheduling regular herd health programs and consultations. Dairymen, too, recognized the value of improving cow health and profitability. In the '70s, having a herd health program was considered the ultimate in improving profits.

Some dairymen resisted these changes and so did many vets. These dairymen liked the idea of eliminating vet costs and saw no value in veterinary consulting services to improve profits. It is likely that many vets who resisted herd health programs saw these dairymen cutting into their professional livelihood. For these reasons, some vets became defensive toward dairymen treating their own milk fever cases, etc. It's

"What do you mean, Bull! When I say I'm ready to start vaccinating, I mean I'm ready to start vaccinating!"

also likely that these vets couldn't "sell" dairymen on the concept of herd health services as an alternative to treating sick cows.

In the '80s, the concept of herd health continued to change in response to the economic and technologic revolution developing in dairy farming. The dairyman and his vet no longer should be satisfied merely with low veterinary costs and healthy cows. The dairyman must squeeze every last penny of profit from his cows. Vets who aren't prepared to supply the service needed for today's production requirements may not remain in the dairy industry. A dairyman needs a vet working with him to develop a program for optimum profits. Likewise, veterinarians need dairymen attuned to programs in production medicine. If the dairyman doesn't make it, his vet won't either.

"And then, this morning, I suddenly noticed she didn't look so good."

While considering the individual components of production medicine, remember that each component is part of the whole program. Dairymen applying individual components may believe they are thoroughly maximizing production, and they may miss the advantages of a total production program. A complete program must include management of the herd rather than just getting cows bred, getting the vaccination done or bringing mastitis under control.

Reproductive Performance

A first component in production medicine is reproduction. If maximum profits/optimum production are to be achieved, a cow must have a calf every year. Research has shown that the calving interval for optimum profits should be 12-12.5 months. Income over feed costs increases when the calving interval narrows from 13 months to 12 months. When the calving interval goes from 13 months to 15 months, income over feed costs decreases greatly.[1] Calving intervals over 13 months cost dairymen $2/cow/day.[2] Cows must be pregnant by 118 days if the calving interval is to be less than 13 months.

Too many dairymen are satisfied with a 13.5- to 14-month calving interval. This is not very profitable. To achieve a calving interval of less than 13 months, a number of factors must be controlled.

First, every cow that freshens should have a reproductive exam 2-4 weeks after calving. Any calving-related problems, such as metritis (uterine infection) or injury to the uterus, should be diagnosed and treated appropriately. "Unsuccessful calving" is a term referring to cows that have had milk fever, retained placenta or other problems at freshening. Cows with an unsuccessful calving should be evaluated on the next regularly scheduled visit by the veterinary consultant, even if more than 2-4 weeks after freshening. This exam should be in addition to either a special sick call visit for her at freshening or intensive therapy given by the dairyman. Cows that have been fresh 45 days and not yet seen in heat should also be checked at regular reproductive exams. Cows that have been inseminated or bred naturally 30 days or more before the visit for reproductive exams should be checked for pregnancy.

Cows found to be pregnant should be recorded. Cows that are not pregnant should be evaluated and it should be determined whether they need to be rebred or are "repeat breeders" that require further evaluation and therapy.

While a few dairy technicians have the skill to diagnose early pregnancy, cows with breeding problems should be evaluated by a vet skilled in the medical, hormonal and physiologic management of cows with breeding failures. Not every vet can do this type of evaluation. If your vet doesn't have the skill, there are 2 options: First, discuss the need for these skills in your herd and give your vet a chance to develop them. Second, if the first alternative won't work, find a veterinarian who can do these examinations.

Cows should be evaluated at regularly scheduled visits for normal fertility and early pregnancy. After a problem breeder has been identified, appropriate treatment must be started. Sometimes, this may not involve anything expensive or time consuming. Other times, therapy may involve much time. The measure of the success of this program is in breeding efficiency, a part of reproductive production medicine. Various factors are a part of this. Interval to first heat and interval to first service should be measured. Average days open is a dynamic tool used every month to monitor the calving interval. Ultimately, the total reproductive management is summed up in services per conception and the calving interval.

The frequency of reproductive evaluations should be determined by the requirements of your herd at its production level. Usually, larger and high-producing herds require more frequent visits than small or low-producing herds.

Disease Prevention

Production medicine involves having every cow on a disease prevention program. Cows should be vaccinated for the diseases prevalent in your area. Generally, nearly every herd should be vaccinated for IBR, BVD and 5-way lepto. Some herds also should be vaccinated against *Hemophilus*. Other vaccination programs are optional and should be tailored according to the herd's needs.

Procedures should be developed for entry of new animals into the herd. Your vet can tell you how these animals should be isolated and what health screening measures are needed. Procedures for management of sick cows in the hospital pens are warranted. Some herds may need a Johne's disease control program. Other purebred herds in the export sales business may need a leukosis control program. Deworming recommendations should be based on the herd's requirements.

Dairymen often observe that their vet's skills grow with their herd's growing health requirements. Many vets develop more skill in management practices as needed to prevent diseases. These skills include knowledge of optimum ventilation, milking machine principles and

"You're too late!"

disease surveillance in herds. Veterinarians have the responsibility of providing expert advice, while the dairyman has the responsibility of listening and weighing that advice. Every piece of advice may not be workable for the situation but serious consideration of these points often creates better programs.

Nutrition

A production medicine program also considers nutrition. This is usually the most important factor in increased production and profit. Some vets are quite expert in offering advice on nutrition. Others rely on qualified nutritionists for advice. In either case, there should be a continuing dialog among the dairyman, veterinarian and nutritionist, including exchange of information on current technology on feed and additives, analysis of feed, and nutritional deficiencies (Fig 1). Contact must also be maintained with feed suppliers.

Nutrition cannot be taken lightly. Too many problems relate to the nutritional status of the herd. Tissue assays of nutrients to ensure adequate levels of minerals are critical. Liver and blood tests for copper, zinc, selenium and other nutrients are important if a herd is to reach optimum performance. Feed costs are a major factor that can be evaluated and controlled.

Udder Health

Another component of a production medicine program is the mastitis control and udder health program. A very important factor in the udder health program is regular surveillance for mastitis. This includes

Figure 1. Testing your forage is an important facet of your nutrition program. Take random core samples of at least 8-10 bales and mix the core samples for a lab analysis of their nutrients.

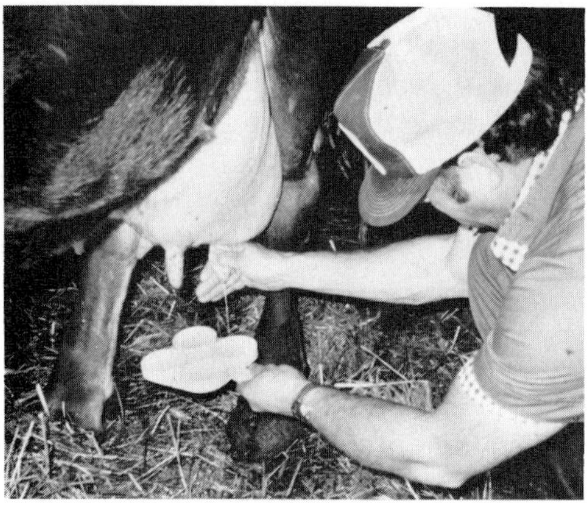

Figure 2. One of the best cowside procedures for detecting a high somatic cell count is the California mastitis test. A paddle is used as a reservoir for milk taken from each quarter. Test solution is added to the milk samples. If the somatic cell count is high, the mixture becomes very thick. The degree of thickness indicates how high the somatic cell count is.

somatic cell counts, bulk milk tank sampling and routine culturing of high-SCC quarters (Fig 2). Also necessary is an analysis of milking machine operators and milking machine function. Milking machine operation analysis involves working with the vet or a representative of the milking machine manufacturer. Occasionally, outside consultants are the source of this expertise. Vacuum stability is important for udder health.

Cows with a high SCC, positive milk cultures or clinical mastitis should be brought to the vet's attention. His or her expertise is invaluable to evaluate production losses from high SCC and other economic factors. Economic factors play an important role in treatment and prevention decisions. Teat dipping, predipping teats, types of dips, and environmental causes of mastitis should be included in the mastitis control program.

Dry-cow treatment should be based on information from the milking herd. Evaluation of dry-cow housing, the calving environment, proper calving assistance, and dry-cow nutrition is essential in production medicine. Dry-cow teat dipping, udder examinations, precalving milking and udder edema evaluation are also essential. These factors should be evaluated for maximum response at calving. Decisions ultimately fall on the dairyman's shoulders. To make proper decisions, he needs all of the input provided by a veterinarian.

Calf and Heifer Management

Probably the most neglected area of production medicine is calf and heifer management. Their housing is often overlooked. Are they raised in old buildings that are not suitable for anything else? Has the housing been designed for replacement calves and heifers? Are calves getting

Table 1. Daily nutrient requirements for growing large-breed heifers.

Body Weight (lb)	Age (wk)	Daily Gain (lb)	Crude Protein (lb)	TDN (lb)
110	3	1.10	0.63	3.20
220	14	1.76	1.06	4.64
330	23	1.76	1.40	6.00
440	32	1.76	1.75	7.34
550	41	1.76	1.88	8.64
660	50	1.76	1.94	9.90
770	64	1.76	2.16	11.30
880	78	1.76	2.50	12.69
990	88	1.76	2.85	14.20
1100	98	1.76	3.25	15.70
1210	110	1.76	3.71	17.46

(Adapted from *Nutrient Requirements of Cattle,* 6th ed. National Academy of Sciences, Washington, DC, 1988)

adequate colostrum immediately after birth? Part of production medicine is to monitor antibody levels by blood test if necessary. The umbilical cord should be dipped effectively on newborn calves. Grain, hay and mineral levels should be monitored (Table 1). Heifers must be grouped according to age and size. Body scoring helps ensure optimum performance from heifers. Breeding age and size, and conception efficiency should also be evaluated.

Deworming

A deworming program is essential. Steps are necessary to ensure that worm control is effective, yet not too expensive and labor intensive.

Employee Training

An often overlooked way to prevent production loss is by detecting disease. Eliminating potential areas for losses may involve training employees to do their duties more effectively. This training might include doing milk progesterone assays, giving vaccines and handling routine calvings. Reviewing with the vet the various treatment procedures and teaching an employee to also handle various techniques are important. AI refresher courses help maintain good conception rates. Awareness of drug residues and withdrawal times for various antibiotics should be included in the mastitis program. Training should be available to use the Delvo test and other antibiotic residue tests.

Dairymen can't afford a vet to administer every injection. Yet they must be prepared to follow instructions concerning procedure, dosage and withdrawal times. It is often efficient for dairymen to handle routine sick cow problems themselves. However, they must have training on proper procedures and medication for the sick cow. The dairyman should work with his vet in developing the skills necessary for improved sick cow care.

Other Aspects

Quality production medicine programs often involve financial counseling, with a willingness to discuss the pros and cons of a herd program with a banker. Many dairymen aren't comfortable with their vet's being aware of their dairy's financial status. Though a vet's financial awareness of production medicine isn't necessary, the vet usually *is* financially aware, and can be a valuable ally when talking to bankers about expansion plans.

Personnel training and management, and financial advice only work after a vet and dairyman have worked together a long time and have an intimate understanding of each other's goals and problems.

The vet should also provide diagnostic services. For that routine case that suddenly isn't routine, such tests as blood counts, urinalyses, tissue tests, cultures and X-rays are very important. While working up the diagnosis on these cases, the vet must give management advice that will help eliminate the new problem in the herd. One goal should be to avoid repeated health problems that have no confirmed diagnosis or guidelines for their elimination.

Advantages of Production Medicine

The advantages of a production medicine program considerably outweigh the disadvantages. First, it will improve the overall health of the herd and reduce injuries, resulting in a lower culling rate. Thorough evaluation of the nutrition program will reduce nutritional problems. In addition, income over feed costs will increase. Reducing the incidence of mastitis will raise profits. Research has shown that when the SCC exceeds 50,000, it causes losses of about 1.5 lb of milk per cow each time the SCC doubles between 50 and 400,000.[3]

A production medicine program will not only raise production levels, but also lower the incidence of culling because of mastitis. Improving reproductive performance shortens the average calving interval, which translates into a longer productive life for a cow. When the calving interval improves, it is because of less average days open. Excessive average days open cost $2/cow/day. Reproduction will improve because of fewer pregnancy failures. Calf mortality also will be lower.

Improved breeding, udder health and feeding translate into a lower culling level. The dairyman can cull more for low production, rather than being forced to cull his best cows because of health problems. Voluntary culling results in elimination of lower producers. If the herd is being expanded, voluntary culling keeps the operation profitable while maintaining production levels. Involuntary culling is reduced because of less disease, and such expenses as feed costs are reduced. Semen costs will be less with improved reproductive efficiency. Heifer replacement costs will be lower with well-grown heifers that have been maintained on excellent programs of nutrition, deworming and vaccination. These well maintained heifers frequently reach their first calving 30-40 days earlier than only adequately maintained heifers.

Though total veterinary costs may increase, incorporating all of the components of production medicine into one package will reduce health costs per hundred weight of milk produced. This is achieved by higher milk production, fewer sick calls and the improved return on the investment in veterinary care.

There are other benefits to a production medicine program. It establishes a health data baseline so that from one year to the next you can compare what production costs were in relation to production. It provides a basis for management decisions. The health of the herd is in the hands of a professional. A dairyman has many areas of expertise to supervise. He often recognizes that he's spread thin and can't cover it all. Veterinarians can assume timely and effective treatment. Also, the veterinarian can be a valuable business reference when it comes time to establish the credit necessary for increasing your herd and facilities.

A Case in Point

Larry was one of those clients who really worked hard at dairying. He got up at 3:30 every morning to go to the barn and milk. He was milking 85 cows through a side opening Surge parlor. It took him 4 1/2 hours to milk. Larry was a progressive dairyman. In the '40s, he was instrumental in getting frozen semen available in his area. He went to all of the dairy meetings. Whenever the lights were turned low to show slides, Larry promptly fell asleep and snored through the meeting. You could see his dedication to his cows, but he was at a disadvantage because he didn't have a veterinarian close by to give him good veterinary service. I had to drive about 25 miles to see Larry and look at his cows. I talked him into a forerunner of a production medicine program.

This was many years ago, before anyone had heard about production medicine. Larry didn't want anything so "radical." But because of a lack of veterinary service and his interest in being certain he would continue to get veterinary service, he agreed to the program.

The program consisted of regular reproductive examinations, advice on mastitis and calf rearing, and some basic management advice. Larry could hardly contain himself because of the veterinary bills he started getting each month. In the space of 3 months, his vet bill averaged $100 a month. That was more than he had spent the whole previous year. One hundred dollars a month doesn't sound like much for vet service, but this occurred 20 years ago when vet services hadn't reached the levels of today's services.

Larry agreed to stick with it for a year. At year's end I was anxious to find out whether I still had a client. I knew Larry had spent a lot more money than usual. But he admitted he'd had fewer sick calls that year. He eventually sat down to do his income tax and discovered that he had spent $1200 for veterinary care but had sold $12,000 more milk from the same bunch of cows with very few changes in the rest of the operation. He didn't need any more convincing to stay on the program, which he did until the day he retired. He never regretted for a minute what it cost for veterinary services. He readily agreed the services cost him money but he was selling more milk than ever before.

If your herd is on a production medicine program and it's not making money, you should re-evaluate the program. You either need a different vet or you need to change some of your own management habits. The production medicine program should mean more milk to sell, the ability to expand the herd or improve the profits from breeding stock sales.

References

1. *J Dairy Science* 184:636.
2. *Illinois Dairy Digest* 15(2), 1986.
3. *Vet Clinics North America,* 1984.

5

Maintaining a Healthy Herd

Disease in a dairy cow may be linked to an infection or condition that develops from something in her environment (Fig 1). Infection can be bacterial in origin, such as *E coli* mastitis. It can easily kill a dairy cow. A second type of infection is from a fungus, such as ringworm. Another type of fungus can infect the udder or cause baby calf diarrhea if antibiotic therapy has been overused. Other infections may be caused by protozoa. Examples of protozoal infections are anaplasmosis and trichomoniasis. Viruses can also cause disease, such as infectious bovine rhinotracheitis (IBR) and bovine virus diarrhea (BVD).

A dairy cow may also become diseased from internal disorders. These conditions may be caused by something in her diet or something in her physical make-up. Examples include ketosis and milk fever. Both of these relate to how a cow is fed. Endocrine imbalances, such as cystic ovaries, can also cause disease. Nutritional imbalances of potassium, calcium and magnesium can cause excessive udder edema. Other conditions are related to organ degeneration from age or injury, such as kidney disease. A cow's kidneys gradually lose function from a number of toxic insults over the years, and eventually begin to fail.

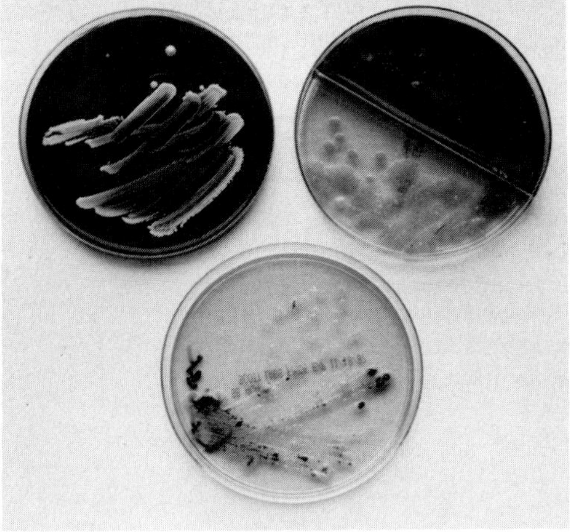

Figure 1. A blood agar plate is used to culture organisms collected from an affected animal. The plate allows organisms to grow and helps a vet identify the types of organisms causing an infection.

A cow can contract disease in a number of ways. One is direct contact with other diseased cattle. Sometimes a cow can contract disease by contact with a disease-laden object, such as a nose lead contaminated with virus particles.

Animals may also become infected and become carriers of the disease without ever showing signs of illness. Cows can carry *Salmonella* without becoming sick and can contaminate the premises, leading to infections in other cows.

Cows contract anthrax from contaminated soil. Contaminated feed and water spread disease, too. You've probably heard that a cow can abort from eating moldy feed, or contract Johne's disease by eating manure-contaminated feed.

Some infections are carried in the air, especially in poorly ventilated barns. A cow with a virus infection can pass it to other cattle just by coughing. Some infections are caused by a bite from an infected blood-sucking insect. Typical examples are mosquitoes carrying either leukosis or anaplasmosis.

Defense Against Disease

The body has several mechanisms to fight diseases. The primary defense mechanism is the skin, which keeps many infections from invading the body. Mucous membranes around the eyes, nostrils and vagina work with the tears and other secretions to wash infections away. There are also specialized cells in the bronchi and windpipe that use tiny hairlike projections to continually sweep infectious organisms out of the lungs. Coughing helps in this process, too.

A cow can swallow things that cause infection. Fortunately, gastric acids in the abomasum (fourth stomach) kill many types of organisms.

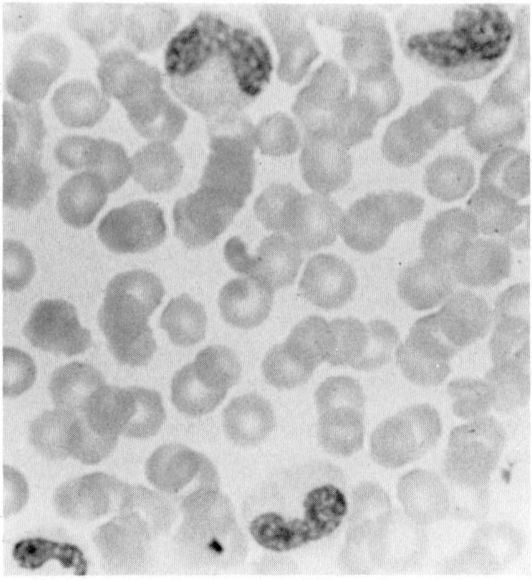

Figure 2. This is a blood smear as viewed through a microscope. The numerous circular shapes are red blood cells. The 4 larger cells scattered throughout the photo are white blood cells.

She can also pass some of these organisms through her body by developing diarrhea, which excretes the infectious organisms. Another defense mechanism that kicks into action if needed is the antibody or immune system. White blood cells and molecular substances help to direct an attack on invading organisms. A cow has an innate capacity, called active immunity, to identify infections. Her white blood cells and antibodies identify diseases and react to them promptly (Fig 2). This active immunity develops after a couple of weeks' exposure to the infection or after vaccination against the infection.

Another type of immunity is called passive immunity. A calf receives it from its dam's colostrum, but it works only temporarily. Passive immunity usually lasts from birth to 4-6 months of age. It gradually tapers off as the calf begins to develop its own active immune system. Passive immunity only develops if the calf has nursed promptly at birth and the cow has quality colostrum. Ideally, the calf must nurse within 6 hours after birth if it's going to receive optimum passive immunity from colostrum. The calf's digestive tract will readily absorb colostral antibody for 6-8 hours. Then there is a gradual decline in absorption over the next 4-6 hours.

Once the calf's active immunity is working, it performs like an adult animal's and can fight infection when the calf has been exposed. Once immunity is developed, it may last a lifetime. In other instances, the immunity may last only 6-12 months. Therefore, some vaccinations must be repeated annually. It's interesting that immunity developed from actual exposure to disease usually lasts longer than immunity from vaccination.

General Signs of Disease

Fever

It's helpful to understand what happens in a cow's body in response to an infectious disease. After an incubation period, signs of illness develop. Cows normally have a temperature of 100-101 F. After incubating an infection, the cow's first response in fighting the infection is a fever. Occasionally this system can run amok. The fever may reach 108 F and continue to rise. In these cases, the cow is likely to die. The infection, or toxins from the infection, causes the fever in these cases. Fever and other signs of illness caused by a toxin are called toxemia.

Fever can also occur because of electrolyte imbalances, blood loss or inflammation of various body systems. Besides acting as a natural defense response, fever also puts the cow under more stress. Because she doesn't feel well, she may not eat or drink, and may become dehydrated. Dehydration increases if she develops diarrhea in an attempt to flush the infection from her body. In addition to losing fluid, a dehydrated cow's electrolyte balance can go out of kilter. With diarrhea she loses the electrolyte called bicarbonate, causing a relative excess of chlorides. Chlorides are acidic, so the cow develops what is referred to as acidosis.

In a converse situation, where chlorides are in short supply and bicarbonate is in excess, the cow has alkalosis. A cow with a displaced abomasum may develop alkalosis. Acidosis or alkalosis can cause serious repercussions. Affected cows may eventually become paralyzed and die if they are not treated.

Changes in the Eyes

Early in a disease process, a cow's eyes may have a sunken and dull look. This is because she begins to lose the water in her body tissues, especially around her eyes. Sometimes a discharge from the eyes and nostrils may indicate a respiratory problem.

Edema

Cows may also show edema, an accumulation of fluid in an affected tissue. This may result from an injury, such as being kicked or butted. It can also be related to how she's been fed. Excessive dietary potassium causes udder and abdominal edema. A third kind of edema shows up when a cow isn't getting enough to eat and is suffering from malnutrition. The edema in these cases usually is under the chin, resulting in the term "bottle jaw" (Fig 3). Other cases of edema can be caused from heart or liver failure and occur in the brisket.

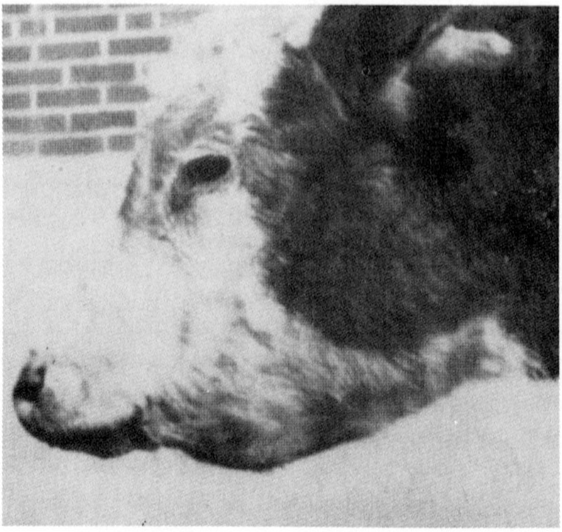

Figure 3. Years ago, bottle jaw was regarded as a disease. Today, we know that a cow with swelling under her chin is probably suffering from protein loss caused by debilitating illness. This cow has chronic diarrhea from long-standing gastrointestinal parasitism.

Changes in Appetite

In addition to the sunken eyes and edema, sick cows may show a loss of appetite. One of the best features of some of the new feeding systems is the daily printout that tells whether each cow ate every pound of feed she was offered. A poor appetite is often an early sign of disease. Some sick cows may eat or drink unusual things, such as dirt, other cows' urine or stagnant water from around the silo.

Changes in Behavior

If a cow is seriously ill, she may begin to behave differently. She may stagger and stumble and become sluggish. She may lie down frequently and not associate with other cows. When she's lying down, she may also grind her teeth. Occasionally, a sick cow is belligerent and hyperactive. She may jump gates or charge people when they try to help her.

In other instances, a cow goes into shock. This is very serious. It can result from an excessive loss of blood, or lowered blood pressure caused by a toxin. A cow in shock is cold and clammy to the touch, particularly on her ears and other extremities. She's often weak and wobbly, short of breath, and reluctant to walk. Sometime she goes down and can't rise. In a case of blood loss, her mucous membranes look pale and she has low blood pressure. It's sometimes hard to find a vein for intravenous treatment. If you don't take immediate measures to deal with shock, the cow will probably die.

Specific Signs of Disease

Udder Problems

No discussion of signs of disease would be complete without discussing the cow's udder. Look for abnormal secretions, odors, heat, pain, swelling or hardness. You can also use the California mastitis test (CMT) to detect inflammation in the udder. A cow's udder may be so painful that she appears lame with the leg that contacts the sore part of her udder. You won't know for sure until you inspect the udder thoroughly.

Uterine Problems

It's easy to miss problems associated with the uterus. Signs to look for are an abnormal discharge, foul odors, or the placenta protruding from the vulva. To detect other uterine signs the cow must be rectally palpated. You and your vet will need to take action if there is a fluid-filled, oversized uterus that feels doughy.

Respiratory Problems

Signs of respiratory disease in a cow include breathing with her abdominal muscles rather than chest muscles. Mouth breathing and panting, and abnormal stances, such as standing wing-shouldered, also are signs of respiratory infection. Some cows may have a raspiness to their breathing.

Digestive Problems

A number of signs may be associated with the digestive tract. A cow may have diarrhea or constipation. When you listen to her rumen with a stethoscope, you may find that it isn't moving. What you normally would hear is a sound like intermittent thunder. Bloat is also a sign of digestive tract problems. Once a cow's gut is greatly distended with gas, she may have breathing problems. In other cases, a cow may have paralysis of the nerves to the stomach so that she appears bloated (or apple-shaped) on the left side and pear-shaped on the right side when viewed from behind. Vets and dairymen sometimes call this a "papple" (Fig 4).

Blood in the stool can be an indication of digestive tract infection, such as BVD, coccidiosis or salmonellosis. In other instances the blood can be from winter dysentery, an ulcer, moldy feed or rough rectal palpation.

Urinary Problems

Cows may develop advanced urinary infections that cause them to pass blood in the urine. Usually blood is only passed when the infection

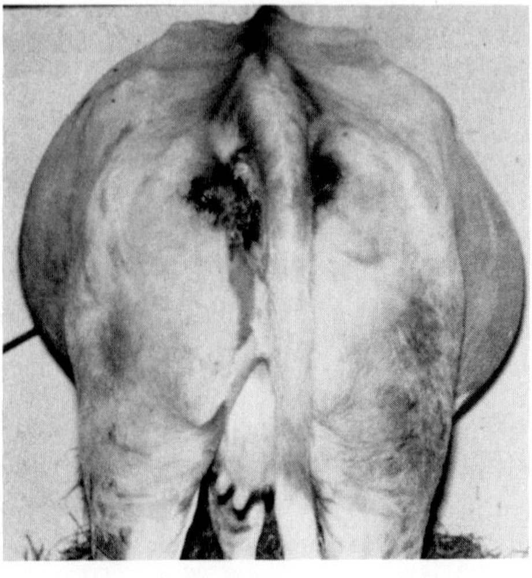

Figure 4. This is a rear view of a heifer with a "papple." Note the apple shape to her left abdominal profile. In contrast, her right side has a pear shape. Thus, the name "papple."

has been going on for a long time. Leptospirosis also can make a cow urinate blood. Cows that have had toxic injuries to their kidneys also pass blood in the urine. You must be careful not to mistake blood passed from the uterus after calving for blood in the urine.

A cow eating less than her energy requirement will pass ketones and acetone in her urine. This occurs when she is ill and not eating enough to supply the energy for milk production, or when a nutritional deficiency prevents her from using all her energy.

Preventing Illness and Injury

Avoid Overcrowding

An easy way to deal with illness and injury in your herd is to give antibiotics. This is often a poor management decision. Because many diseases do not respond to antibiotics, they are managed much more effectively by reducing stress and eliminating sources of disease.

First, become aware of stress that's caused by high herd density. The more cows you shove into a facility, the more stress the cattle feel. Building and lot design is important. For example, cows experience stress if they're confined to an area where manure accumulates. They need access to an exercise area. Other stresses include a parlor with high entry steps or short parlor stalls, which cause cows to develop abscesses on their pin bones. The worst kind of barnlot design sends cows through stagnant water on their walk to the parlor. They may even drag their udders in the water as they move through it. The success of the herd depends on sanitation, so it's important to make a habit of removing manure and providing fresh bedding regularly.

More is Less

Years ago, a dairyman named Leslie had a terrific herd of Holsteins. His cows were milking 16,000 lb when it was common for good herds to have only 12,000-lb rolling herd averages. Leslie fell short on many things, though. He was milking 80 cows in a barn that was designed for about 50. Leslie's barn was short on bunk space, too. The herd strategy there was survival of the fittest, because if a cow wasn't tough, she was going to get crowded out and wouldn't milk well enough to stay in the herd. His cows had breeding problems and poor feet. Leslie also had to deal with a number of milk fever cases because his dry cows ran with the milking herd.

He had an average culling rate of 35%, but Leslie always managed to have heifers coming on to take over.

When he retired, Leslie dispersed his entire herd. What happened to his 16,000-lb cows when they went to other herds is the whole point of my story. Most of these herds weren't overcrowded. They had adequate feed, bunk space and water, and herd managers who did a good job of cleaning up the manure. Leslie always thought he was doing well with his cows. But when they moved on to new homes, they milked their hearts out. Some gave 25,000 lb and 26,000 lb. It's something to think about when you find yourself trying to squeeze in an extra 10 cows to keep the cash flowing.

Control Flies

Your concern about the welfare of your herd isn't complete until you take control of another factor: flies. These include houseflies, hornflies and faceflies. All start pestering cattle in the spring, but invariably dairymen wait until July to ask what they can do to control the swarms on their farms. Fly control must start when the first flies hatch out and begin to lay eggs.

Early in the spring, manure control and disposal are essential. It's a good time to put insecticides around for control. Some dairymen use fluorescent electronic "zappers." Fly control isn't something you can accomplish just by spraying flies. Your best bet is to make a multi-faceted attack, and your reward will be less disease and mastitis.

The key to controlling houseflies is to understand that they have tremendous reproductive potential, and a life cycle of less than 10 days. Adults are fairly long-lived, and the larval form (maggots) develop in bedding and manure. Center your control around sanitation. You can destroy large numbers of maggots in bedding by cleaning out calf pens weekly. Studies indicate that the major source of flies are calf pens, followed by heifer pens and, finally, the manure pile out back. No amount of chemicals or number of electronic zappers will match the positive effect of manure cleanout, so that is the first line of defense.

You can also buy parasites that feed on houseflies. The strategy is to release them in the barn at weekly intervals. It's an excellent idea but not yet very feasible, because these helpful parasites are not sufficiently competitive to the housefly.

As a last resort, you can eliminate some adults with insecticides. Be prepared to change classes of the chemicals with every treatment because flies develop resistance rapidly. For example, follow the use of an organophosphate insecticide with a pyrethroid type.

Another pest is the hornfly. It's about half the size of the housefly and is usually found on the backsides and bellies of pastured animals throughout the summer. Males and females suck blood from cattle and lay their eggs in fresh cow patties. Over the winter, the flies rest in the adjacent soil in the form of pupae (or larvae).

Some researchers say you can improve milk production by 1-2% and increase heifer growth by 3% by controlling hornflies. Dairymen control the flies with insecticides in ear tags, dust bags, oilers, sprays and tail tags. The most popular method is to attach ear tags impregnated with 10% insecticide, which should provide 99% hornfly control.

The third problem fly is the facefly. It looks the same as a housefly, but is found around the nasal and eye secretions of cattle out on summer pasture. Larvae develop in fresh cow patties that are less than 5 hours old. Adult flies feed only in the daytime. They will not enter the barn in the summer, though they winter in barns and houses as adults. Faceflies are highly mobile and move from herd to herd.

The economics of controlling faceflies are somewhat questionable until you consider that they carry the eye worm and pinkeye. Consequently, it's important to control faceflies if you want to eliminate pinkeye. Because hornflies and faceflies have similar breeding habits and attack animals on pasture, control for faceflies is the same as for hornflies. The insecticide dose for faceflies is higher because the facefly is larger and is not on the animal as long as the hornfly. This means dairymen usually put 2 ear tags on cows exposed to faceflies. Faceflies spread so readily that the best control will only achieve about a 60-80% reduction in numbers.

Control Other Pests

Another pest problem is lice. There are 2 types: biting and sucking. Biting lice feed on skin debris and are cream colored. Sucking lice feed on blood and are gray-blue. Lice spend their entire life on a particular species of animal. They are spread by direct contact and reach maximum numbers in the winter. One way to limit louse numbers is to have well-fed animals on a high nutritional plane. We used to think that lice caused anemia and made animals sick. Now we know that sick animals actually allow a large louse population to develop.

Most lice on dairy cattle are the biting kind. Insecticides are the only treatment for lice-infested animals. Because of the potential for residues in milk, you must use different treatments for milking cows, heifers and dry cows. For milking cows, a number of topical (applied to skin) formulations are available, whereas systemic (fed or injected) insecticides can be used on nonlactating cows. A pour-on insecticide can be used on lactating animals. Cows should be retreated in 10-14 days to completely eliminate the louse population. Once your herd is louse free, treat any animal in contact with your cattle so you prevent reintroduction of lice.

Cattle grubs also are a problem in dairy animals. The adult grub fly attacks animals throughout the summer and deposits eggs on the hair. The eggs hatch into larvae that burrow down through the skin and migrate into the underlying tissue, where they remain for 4-9 months. They eventually emerge and increase in size along the back ("warbles"). Older animals are resistant to these migrating larvae, but young animals are susceptible. Heifers always seem to have the highest grub populations.

Pour-on, spot-on or injectable insecticides work very effectively on cattle grubs. You can't use systemic insecticides on milk cows because of milk residues. If you fail to treat replacement heifers, you'll probably lose the greatest amount of money.

Control Ventilation

In addition to surface contamination, another threat to your herd is airborne contamination. This is a grave problem in barns that are overcrowded and underventilated. Young stock housed in poorly ventilated buildings usually have chronic respiratory disease. Design your buildings to provide natural ventilation. Exhaust air through a flue and draw fresh air in through several inlets. If you don't have hip-roofed barns where this system will work, install mechanical ventilation.

As a general rule, forced-ventilation barns should have at least 4 air changes per hour in the worst of winter weather and up to 30-60 air changes per hour in summer weather. The exhausted air should move through one outlet, but you'll need inlets installed uniformly around the building to provide adequate distribution of the fresh air.

The most profound change in tie-stall dairy barn ventilation in the last 15 years has been the general reduction in barn temperature. With a change no more complicated than a simple turn of the thermostat dial, dairymen have significantly reduced the moisture, gas and microorganism load (Fig 5). In years gone by producers kept the barn at 55 F because it was a comfortable environment to work in. With the recent interest in and emphasis on a healthier barn environment, many dairymen keep their barns at 45 F.

The main change in fans has been the recent popularity of European fans. Most enthusiasts think these fans are more energy efficient than the domestic ones. That is, they may move more cubic feet of air per minute (CFM) with less energy for comparable size. But research

Figure 5. Ventilation rate for a 50-cow dairy barn held at 55 F and 45 F. Note that the cooler barn temperature of 45 F is providing 4 air changes per hour at an outside temperature of 5 F. When the barn temperature is kept at 55 F, the warmer barn has slipped below 4 air changes per hour at 17 F outside. At wintertime temperatures, the ventilation rate is significantly better with the cooler inside temperature. This reduces gases and the microorganism load.

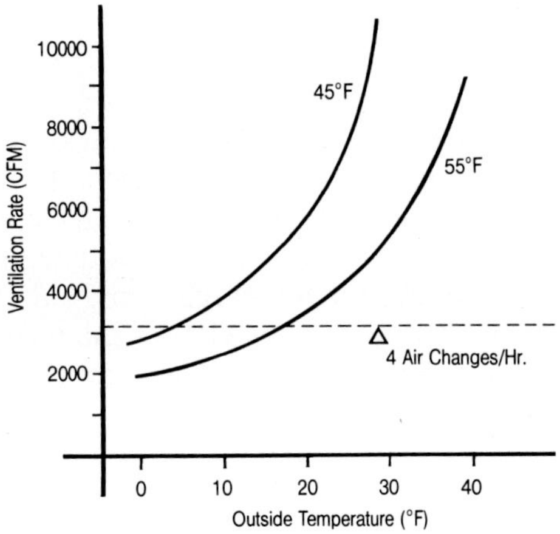

suggests that this may not be the case. European fans have only a modest improvement in energy efficiency. It would take 7-8 years to pay for the difference in initial cost if the fan operated at high speed continuously.

Air inlets for cross-ventilation systems are far more important than fan location in determining air patterns in a barn. *Inlets must be as nearly continuous as possible around the barn to achieve good distribution.* The size of the inlet opening can be adjusted, depending on the number of fans operating. Under most circumstances, cross-ventilation is preferred for inletting and distributing air into a barn. It's simple, inexpensive and, properly done, distributes fresh air without creating a draft.

Good circulation is hard to achieve in some circumstances. Old 2-story barns with low ceilings and one or more walls without any inlets can have dead, stale areas. Many times a single-story addition has much tighter construction than the old 2-story barn to which it is attached. Fans operating in the new wing draw the air not from the inlets, but from the old barn. So the air in the old barn becomes cool and fresh, while the new addition is too hot and stuffy.

Another challenge is the calf room, especially if it has a low winter ventilation rate and supplemental heat. Dairymen complain that there are drafts from the inlets and poor distribution of heat from floor to ceiling. One solution is to buy a polyethylene tube and install it along the length of the ceiling. A fan at one end will move the air, and holes along the tube distribute it throughout the barn. A variation of this tube is a plywood duct with round holes or rectangular slots along its length. The plywood ducts cost more, but they are far more attractive than the plastic tubes. You can also build them wide and shallow so you don't sacrifice head room.

Recirculating air systems have had a bad reputation, and most of this has been deserved. The major complaints are that they cause too high an air speed at the livestock level or that they move too much air. Some dairymen observed that air coming from the holes in the tube hit a beam or collided with an air stream from a neighboring tube and fell to the floor as a draft. Now we know that restricting the recirculation air to 1.5 cubic feet per minute per square foot and avoiding conflict with neighboring air streams, beams and other obstructions will usually solve the problems.

A more fundamental criticism of recirculation is that you end up recycling and distributing microorganisms from sick animals throughout the barn. No matter what the ventilation system, you still need to exhaust air continuously. The best way to reduce the microorganism and gas load, and thereby improve air quality, is to step up exhaust rates.

Heat exchangers have become popular energy-saving devices for livestock buildings, especially in cold areas of the country. But they're controversial, and your payback on this kind of device can be quite variable. The best payback is about 27 months and the worst payback is about 10 years. Factors that shorten the payback period are a high desired room temperature, high exchange rate and/or consistently low outside temperature. Remember that heat exchangers need periodic cleaning and maintenance, too.

Every so often, we find merit in reviving an old idea. One example is ventilating tie-stall barns naturally (Fig 6).

Dairymen like several features of a naturally ventilated barn. First, there's no fan noise. Second, you have enhanced summer ventilation because big sidewall openings let in more fresh air. Third, you have lower maintenance costs because you don't have to deal with fans and controls. And fourth, you spend less on electricity.

There are a couple of thumb rules to follow when you design a building for natural ventilation. The area of inlet opening should be about 8% of the floor area. The area of outlet opening should be about

Figure 6. King ventilation system.

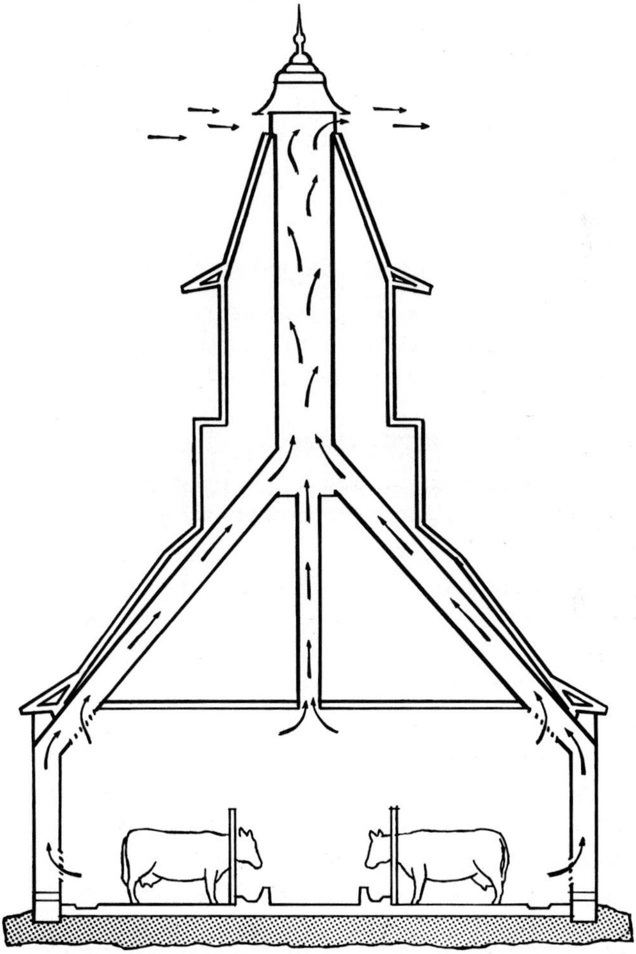

(From *Physics of Agriculture,* 1907)

2% of the floor area. As you might guess, it's easiest to meet these specifications if you're building a brand new facility. You're in for a bigger challenge when you try to refit old barns by installing air inlets and outlets.

Keep in mind that it's far more important to maintain the continuity of the inlet than it is to maintain the continuity of the exhaust. Inlets must be located in direct line contact, whereas the outlet can have several bends. Chimneys are particularly useful in ventilating old barns, because they are cheap and easy to install. Insulate them to an R factor of 10 and extend them above the peak of the barn for good draw. A rule of thumb is that a 4-ft x 4-ft chimney will service about 1,500 square feet of barn area.

You need at least a 10-ft flat ceiling in a barn you intend to ventilate naturally. In old barns with lower ceilings, the ventilation is not likely to be as good. However, the 10-ft guideline also applies if you want to use a fan to ventilate a barn. It's your choice whether you want to control the exhaust and intake openings manually or automatically. Those who choose manual operation like the simplicity, safety, low cost and easy maintenance. But they have to resign themselves to spend time in the barn adjusting things so there's adequate ventilation.

Automation in the form of pneumatically or electrically controlled doors and peak closures reduces the adjustments you must make. However, it adds greatly to cost, complexity and maintenance.

Control Social Interaction

The stress on dairy cows in overcrowded conditions also affects their social interaction. In all groups of animals, there is a "pecking order." The boss cow has free access to food, water, shade and housing, but the timid cow on the bottom end is not likely to get her share. Sometimes that timid cow may be your very best cow. Her chances of reaching her potential are increased if you supply adequate feed bunk space, shade and housing.

Minimize Injury and Illness

Various sources of injury and illness exist under everyday herd conditions. Uneven floors cause slipping. Concrete floors that have been scraped frequently eventually get so slippery that cows may fall and hurt themselves. Uneven high steps into the parlor may cause cows to bang their udders as they enter or slip on steep inclines exiting the parlor.

Cows also may encounter toxic chemicals in their environment. Toxic plants and shrubs, such as wild cherry or snake root, make cows sick. Ornamental plant trimmings thrown over the fence to the cows are highly toxic. Trimmed and wilted rhubarb leaves are deadly to cows.

Don't leave your farm machinery and equipment in the cow lot. Machinery may have grease or oil on it that's poisonous to cows. Cows can tramp over the machinery and lacerate their udders. You might be tempted to leave the corn planter in a spot where the cows could get around it. But the insecticides inside, such as Furadan or Counter, are deadly. One lick of these insecticides can kill a cow.

Your Veterinarian Can Help

Even if you are a top manager who knows how to reduce density stress on your cattle, build for proper ventilation, work daily to eliminate injury and illness, and feed for top production, you still occasionally need the assistance of a veterinarian. You and your veterinarian should work together as a team. The veterinarian should work toward increasing your profits. Likewise, you must realize that if the vet doesn't get a reasonable fee for services, he or she won't stay in business.

The best way to start out with a new vet is to call him or her in for a visit *before* you have an emergency. Find out what kinds of services are offered, and ask for a fee schedule. This is also the time to discuss what kind of professional help you want.

You'll need assistance in 3 kinds of cases: urgent emergencies, less urgent emergencies, and routine non-emergency calls.

Most Urgent Emergencies

- Acute bloat
- Trying to calve for more than 4 hours
- Milk fever in which the cow is lying flat on her side
- Severe hemorrhage from the milk vein or other large vessel
- Nervous ketosis, in which the cow may go berserk or act belligerent
- Major or compound fracture
- Toxemia (grain overload or poison)
- Coliform mastitis (cold, in shock, staggering and deathly ill)

Less Urgent Emergencies

- Milk fever in which the cow is still standing or just went down and is still bright and alert
- Trying to calve for less than 4 hours but the calf is presented abnormally
- Chronic bloat with a slow onset
- Coliform mastitis (but the cow is not severely ill and the milk is watery)
- Lacerations that aren't hemorrhaging profusely

Routine Non-Emergency Calls

- Ketosis
- Retained placenta
- Routine mastitis
- Displaced abomasum

Chapter 5

Before the Vet Arrives

You can do certain things before the vet arrives to help your injured or sick cow and facilitate treatment. The vet will give you specific suggestions if you ask for them. Most important, provide first aid. For example, if you're faced with a hemorrhage or major laceration, a bandage to stop blood flow will do. If there's mastitis in a quarter, strip it and take the cow's temperature. Confine the affected cow in an easily accessible stall or stanchion. There's probably nothing more frustrating for the vet on a busy day than to show up and find you haven't even brought the sick cow in from the pasture. Try to anticipate your vet's needs.

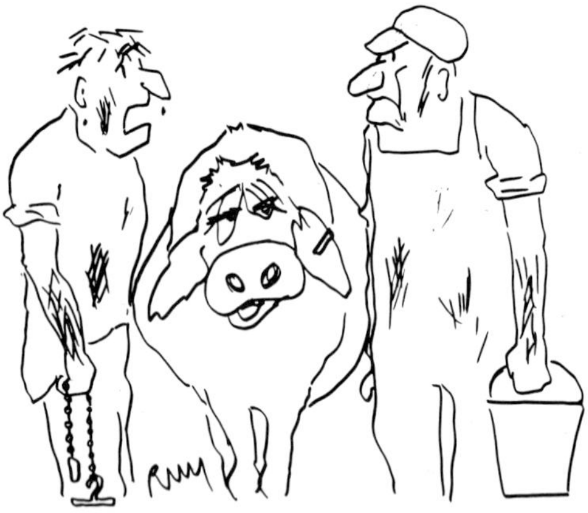

"Well, Dad, it's 2:00 AM and we've been messing with her all night. It's time to call the vet."

Bring out the cow's health records that include the past history of problems. Other necessary equipment includes a bucket of hot water, a halter and/or a nose lead. A headgate is a godsend when you must give medication intravenously.

When the vet arrives, plan to be available to help hold the cow, answer questions and receive instructions. If you can't be there, assign an employee to the task or, at the very least, be sure to leave complete instructions and information for the vet.

I often tell dairymen to "milk their veterinarians for all they're worth." You're paying for time and expertise, so talk freely about your herd. Ask questions and persist until you get explanations. It pays to develop a comfortable but businesslike relationship with this professional.

Sometimes you may feel uncomfortable with the vet you've chosen. If so, look around for another vet whom you feel you can talk to. You'll want to work as a team on vaccination programs, deworming, dehorning and other health care. Sometimes it's best to consider your vet the supervisor. You carry out the plan.

Certain equipment and medical supplies are good to have on hand in a dairy barn:

Medical Supplies and Treatment Facilities

- Treatment/maternity pen
- Headgate
- Bright overhead lights
- Electrical outlet
- Alcohol
- Iodine solution for dipping navels
- Bloat remedy
- Teat ointments for various abrasions and lacerations
- Adhesive tape for bandages
- Sterile dressing material
- Scissors
- Obstetrical chains and handles stored in an anti-rust, general-use disinfectant solution
- Lubricant
- Trocar and cannula for severe bloat
- Clean, disposable syringes *with guards* to cover needles
- Antibiotics for routine infections
- Rectal thermometer
- Flashlight to check a laboring cow at night or to check her eyes for milk fever
- Palpation sleeves
- Roll of cotton
- Paper towels
- Balling gun or 2-ft section of 1-inch plastic pipe
- California mastitis test kit
- Antibiotic drug residue test kit (Delvo)
- Esophageal feeder
- Stethoscope to listen to rumen
- Keto-Stix for testing urine for ketosis
- With vet's okay, intravenous set and needle (stored in disinfectant) for treating milk fever
- Frick tube (stainless-steel tube) to put in the cow's mouth to keep her from chewing a stomach tube
- Stomach tube

Purchase supplies only from reliable sources. Beware of the wholesale salesman peddling unknown brands of vaccines. The product may be nothing more than "snake oil." This is especially true when it is an intrastate (within the state) product rather than an interstate (across state lines) product. The FDA and USDA regulate only interstate products.

Always use medications according to the directions on the container unless otherwise directed by your veterinarian. Discard outdated medications. I also suggest that you only use more than one drug in a cow when your vet advises you to do so. Don't pump your cows full of 4 or 5 different antibiotics.

"Naw Man! I ain't never worked in no animal hospital, but I had lots of experience with drugs!"

With training from your vet, you can handle several common emergencies until he or she arrives. For example, with hemorrhage from a milk vein or teats, you can apply vise grips or forceps to shut off the hemorrhage temporarily. In some cases you may need to give intravenous fluid if much blood has been lost and your vet has taught you the proper procedure.

In cases of bloat, many veterinarians often encourage their clients to pass a stomach tube on cows that are staggering around and having difficulty breathing. Otherwise, the cow may die before your vet can get there. You may need to "stick" (puncture the rumen with a knife) a bloated cow to save her life if she's going down, can't get up, or has stopped breathing altogether. In this case, use a trocar or pen knife, and stab her only on the *left* side about 3 inches in front of the hip bone. Never stick a cow on the right side because the rumen is located on the left side.

Poisoning cases are especially tough. You and your vet should talk about what kinds of antidotes you should have on hand. In most cases, you'll need to turn this emergency over to the vet for proper treatment.

Another relatively common emergency condition is a fracture. You may deliver an abnormally presented calf and discover it has a fracture. Use a splint to immobilize the leg until your vet can set it. Cows that fracture their hip or femur have a poor prognosis. It takes medical heroics to salvage them and you usually have to send them to a university or large animal clinic. Generally, fractures are hard to treat if they're above the hock on the hind leg or above the knee on the front leg.

The other common emergency you may be confronted with is shock. Usually it's best to wait for the vet to treat this condition. He may need to stop bleeding and give intravenous fluid or other appropriate drugs, depending on the cause of shock.

References

1. *Proc AABP, 18th Annual Conference,* 1985. pp 67-68.
2. *Proc Ecol* 3:237-244, 1981.
3. *Minnesota Extension Bulletin,* 1978.
4. *Proc AABP, 18th Annual Conference,* 1985. pp 60-67.

6

Managing Dry Cows and Preparing for Freshening

The dry period in late pregnancy is very important for a cow. It's a time to restore her energy, tissue mineral levels and other reserves necessary for good health. This is the time to rejuvenate her so she can freshen and milk to optimum level. Feeding a dry cow correctly is covered completely in Chapter 19. In this chapter, I'll highlight important general principles.

The way you manage the dry period directly affects a cow's health and subsequent lactation. Don't put dry cows at the bottom of your priority list. Check them regularly for mastitis and control the length of the dry period. A cow with a short dry period doesn't have enough time to replenish her body's reserves, so she may give less milk. On the other hand, a cow that is dry too long tends to get too fat and flabby. She'll probably have trouble calving.

Aim for dry periods of 60 days. Research shows that dry periods of 130-170 days result in 12% less milk.[1]

When it's time to turn a cow dry, there are a couple of methods you can use. One method involves milking the cow every other milking or every other day for several days until her production drops off. Many cows milking at lower levels dry up very effectively when you milk them this way. Heavy milkers don't respond as well.

The second method dries cows off abruptly — a kind of "cold turkey" treatment. Shut the cow in a pen, feed her poor-quality hay or straw, and do not offer water for 24 hours. (In extremely hot weather, though, you must give her a small amount of water.) This method works well for drying up the most persistent heavy milker. At the end of 24 hours, continue to offer low-quality forage but allow as much water as desired. Milk her out once more, then dry her off.

Start a Mastitis Prevention Program

Treating dry cows for mastitis is the most effective part of a mastitis control program. When you treat a dry cow, you don't have to worry about milk residues because she's not lactating. And the cure rate of treating chronic mastitis during a cow's dry period is much higher than during the lactation period. Studies show that treating lactating cows for mastitis may not be economical.

The best time to treat a mastitis problem is when the cow is close to drying off. Turn her dry using a dry-cow preparation to treat the infection. When you use these products regularly on your dry cows, you reduce the incidence of new contagious infections, and give damaged tissue a chance to regenerate before freshening. You've probably seen this process go on naturally in your herd. A cow that goes dry in one quarter because of mastitis may come back in the next lactation milking fine from that damaged quarter.

Here's the best way to administer the dry-cow preparation. First, milk the cow a final time. Next, clean and dry all 4 teats thoroughly. Scrub them with a 70% alcohol pad for 10-15 seconds, and let the alcohol dry (Fig 1). Choreograph your sanitizing so you scrub the teats closest to you last. Don't brush them with your sleeve while reaching for the teats on the other side of the udder. Next, infuse the closest quarters with dry-cow treatment, then the quarters farthest away. Insert only the tip of the tube. Treat all quarters of every cow.

Figure 1. The teat end must be thoroughly disinfected with alcohol before treating for mastitis. Because most infections start through the teat end, you may unwittingly introduce an organism into the udder if you do not disinfect first. Only the very tip of the infusion tube should be put through the teat sphincter. Further penetration into the canal can worsen the infection. (Courtesy of The Upjohn Co)

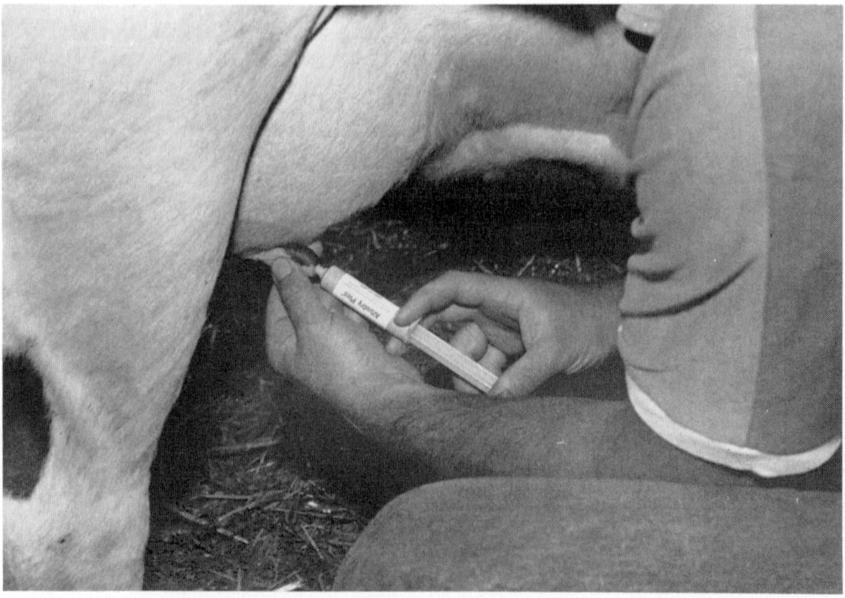

It's wise to stick with single-dose infusions. The danger in drawing multiple doses from one bottle is that you can contaminate the whole bottle of antibiotic after you've infused an infected quarter. You end up contaminating the rest of the udder. Use an antibiotic with a slow-release base so you get prolonged action of the antibiotic. Avoid lactation-approved mastitis tubes as dry treatment because they only work for 2-24 hours. Sometimes you may lose faith in your dry-cow treatment program if your springing cows come down with mastitis. There are several reasons why this can happen. Some infectious organisms, such as *Staph aureus* and *E coli*, are unaffected by dry-cow antibiotics. The dry-cow tubes are formulated to fight *Strep* and some *Staph*, and have little effect on *E coli*. Also, environmental pathogens, such as *E coli*, challenge the udder throughout the dry period, often after the dry treatment is no longer acting.

You may know dairymen who fight mastitis without dry treating every cow. There are some alternatives, and they have certain advantages and disadvantages when you compare them to treating every dry cow. One method is to culture each cow's quarters and only treat infected quarters. You base your treatment on the culture and sensitivity results from the laboratory. You spend less on antibiotics and can use antibiotics known to fight the specific infections in your cows.

Illinois research indicates this is a valid strategy in eliminating existing infections.[2] But it has an economic drawback: Cultures are often more expensive than dry-cow treatments and cultures may report false positives and false negatives. Some dairymen treat dry cows selectively, according to somatic cell counts (SCC).

In general, culturing and treating infected quarters is not as effective in controlling mastitis as treating every cow with dry-cow treatment unless you can be sure that your dry cows spend their nonmilking days in a very hygienic environment. Selective dry-cow treatment may fail when therapy is based on past history of infection. Some cows will be treated that don't need it. Some cows get new infections because they weren't treated. And this method tosses out the whole theory of preventive care, which really ought to be your basis for management.

Fighting mastitis in dry cows doesn't stop with treatment. Continue to put your dry cows through the parlor and teat dip for several days into their dry period because they are more vulnerable during this period. Various teat dips are available. The routine disinfectants don't prevent the environmental infections, though. Some dairymen believe the dips that form an acrylic or latex seal over the teat end do an excellent job of fending off environmental mastitis. They are moderately effective immediately after you apply the dip, but not for the whole dry period.

Look Closely at Minerals and Vitamins

One of the most important minerals for a dry cow is calcium. She needs enough for her own maintenance plus whatever her growing fetus requires. A large-breed cow needs 60-100 grams a day, while a small-breed cow needs 50-60 grams a day. Excessive calcium will increase the incidence of milk fever. A fresh cow must quickly draw on her calcium stores when she delivers her calf. If her body can't respond correctly, she "crashes."

Also watch selenium and vitamin E closely. You can't talk about one without talking about the other. Too often, people think a deficiency in either of these nutrients is strictly an Ohio problem. In fact, there are areas in about 44 states where selenium/vitamin E deficiencies are a major problem for dairy herds. Experts are beginning to recognize that a deficiency of either during the dry period is a major cause of cystic ovaries in fresh cows, and low levels impair the cow's overall health. If a serious deficiency exists, it doesn't matter how well a cow is fed. She'll produce poorly and have low resistance to disease.

Give Dry Cows Enough Exercise

Cows that aren't milking need exercise to maintain muscle tone. They must exercise their skeletal muscles and the muscles of their digestive tract. One way to ensure activity for their skeletal muscles is to design your dry-cow housing so the cattle can come and go as they please. Cows that are crowded in box stalls exhibit muscular weakness at calving.

An On-Farm Example

Terry quit dairying a couple of years ago. When he had cows, he worked hard to give them excellent care. He was an intense, handsome guy who could knock women off their feet, but his cows always came first. One time, he even took a blind date in her best dress to the farm so that he could calve a cow out at 12:30 AM. He never did take the lady to dinner that night, but she must not have minded, because she eventually married him.

Terry had an Elevation daughter that the classifier scored 91. She had milked nearly 30,000 lb in the last lactation so Terry had high hopes for her next lactation. Not only was she bred to Valiant, but he thought she could win the individual cow milk record in the county. Terry kept her in a box stall during her dry period. The cow tended to be heavy but he watched her ration closely. She freshened with a bull calf and went down with milk fever. Terry was beside himself with worry. He had her treated and she got up. The next morning she went down again and never did get up. After a week of nursing care, he finally put her to sleep. The cow may have died regardless, but her lack of exercise during the dry period certainly stacked the odds against her.

Another way to promote exercise is to plan for it in the way you feed your cows. For instance, you can locate the feeding bunks and the water troughs at opposite ends of the exercise area. To eat and drink, the cows must walk the length of the barnyard. Or you can situate the feed and water some distance from housing.

During the dry period it's important to stimulate the smooth muscle tone of the digestive tract, as well as the muscles of the skeletal system. If the smooth muscles of the intestinal tract become flabby, a cow is prone to digestive problems, such as displaced abomasum, after calving. To develop tone during the dry period, feed forages that require the intestines to work to digest them. A dry cow eats about half as much feed as a fresh cow, so her forages must be twice as resistant to digestion to maintain smooth muscle tone in the digestive system. Good dry-cow forages are high in fiber. The best sources are long-stem dry hay, corn stover and oat straw. If you're going to do limit feeding, feed some corn silage, and you may want to add oat straw or ground corn cobs to increase the fiber level.

Develop a Critical Eye for Condition

It's important to learn to recognize when dry cows are getting too fat or too thin and then adjust rations accordingly (Fig 2). This "critical eye" does not come easily to many dairymen. It may take years to develop the skill. Even with practice you may fall short. If that's the case, couple your observations with advice from your vet or another cattleman with a good eye.

Aim to have cows go into their dry period in good flesh and then help them maintain that flesh until they freshen. You have a special challenge when you're faced with dry cows that are too fat to start off with (Fig 3). It's too late to take the extra weight off by reducing their feed intake; you can limit the amount of high-energy feed for exceptionally fat cows. In the last 2-3 weeks before freshening, you can also feed niacin to help mobilize the fat from the liver and stimulate their appetite.

Figure 2. A dairyman must be able to recognize individual cows that are not carrying the flesh they need for optimum milk production, and adjust their feed intake accordingly.

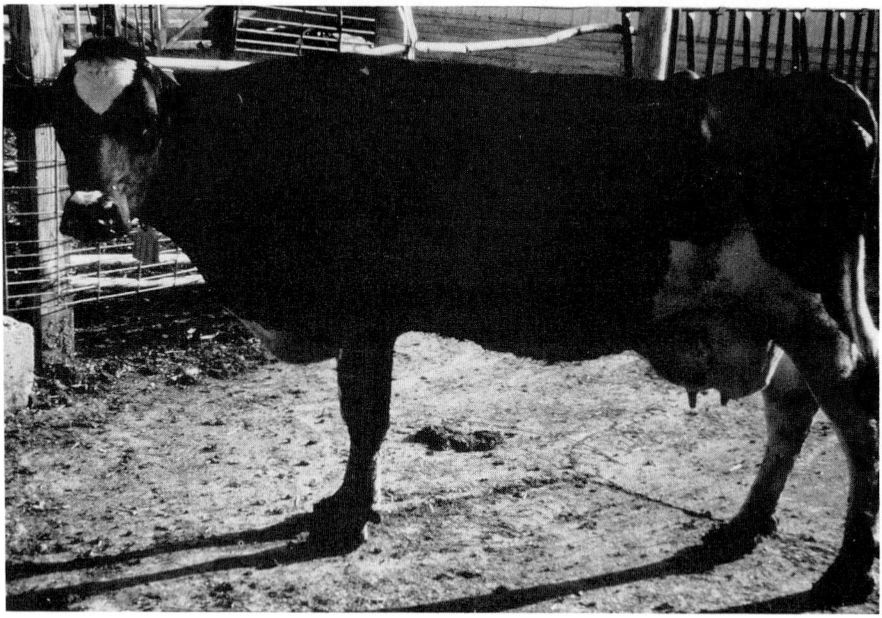

Figure 3. This dry cow is too fat. Note that her pins and hooks are covered with fat. Fat cows are likely to have numerous metabolic problems after freshening.

Special Care Before Freshening

There are a number of things you can do for the dry cow to help her prepare for calving. I call this "prefreshening care."

Vaccinations

Guard your cows against infectious bovine rhinotracheitis (IBR). You and your vet can use either a killed injectable vaccine or an intranasal vaccine. A new live IBR vaccine product has been chemically altered so it's safe for the pregnant cow. Under no circumstances should you use the conventional modified-live IBR vaccine on pregnant cows. It may cause abortions.

The second disease to guard against is bovine virus diarrhea (BVD). It's best to use a killed vaccine on dry cows. The conventional modified-live BVD vaccines are not safe for pregnant cows. New evidence indicates that live BVD vaccine may not be safe even in open cows. It may perpetuate virus carriers that can spread BVD virus. Vaccinate your dry cows only with killed BVD vaccine or consult your veterinarian regarding the alternatives.

The third vaccination recommended for dry cows fights 5 types of leptospirosis, a disease that causes infertility, abortions and mastitis in dairy cattle. There are several optional vaccines you may want to ask your vet about. These include *Hemophilus somnus, Salmonella, Clostridium* and *E coli*. You may wish to vaccinate dry cows against *Hemophilus somnus*. This vaccine isn't a "must" for every herd. Talk with your veterinarian about it.

Salmonella is a major cause of diarrhea and subsequent death in newly freshened cows. Some farms have a chronic salmonellosis problem. Since there are only a few commercial *Salmonella* vaccines available, you may need to ask your vet to arrange for one to be prepared. But you don't need to give this vaccine if you've never had salmonellosis in your herd. Vets usually use a 7-way vaccine for *Clostridium*. Fresh cows that get this organism in the uterus develop a very serious uterine infection. Vaccination against *E coli* is not intended to help the cow, but rather the calf. The vaccine is given to raise antibody levels for the newborn calf via the colostrum.

Deworming

There's a lot of debate over whether dry cows should be dewormed. In most cases, it's not necessary unless they are on pasture. You may not realize an economic benefit unless a dry cow has some immune problem that tends to make her more susceptible to parasitism. If that's the case, various dewormers are available. You can add deworming agents to the feed. There are also injectable, paste or bolus dewormers available. As with any other medication, follow the package directions and observe any withdrawal times.

Set Up a Calving Area

There's no substitute for a calving area that's separate from the rest of the cattle housing. Freestalls aren't an adequate environment for calving. The free stall bed isn't very clean, and the cow may end up dumping her calf out in the manure of the aisleway, where it's slippery and dirty. She may get trapped under the edge of the freestall and not be able to get up. If she's able to get up and out of the stall, she's likely to do the "splits" on the concrete, because she may be weak and wobbly on her hind legs. These conditions are also ideal for her to prolapse her uterus.

The best calving environment is a small, dry, grassy lot that's free of contamination, where cows can calve in relative comfort and safety. Don't use this lot for other cattle. That way you reduce exposure to microorganisms that can cause disease in fresh cows and their calves. Both cow and calf are vulnerable at freshening.

In large herds, especially in the southwestern United States, small, dry, grassy lots may not be available. At these dairies, a dirt lot reserved for freshening cows will work fine. When a cow has freshened and the lot has gotten messy, scrape away the top layer of soil and put down fresh clay before the next freshening.

In most of the United States, winter weather discourages outside calving. So give up the grassy lot for the season and move freshening cows inside to box stalls. Scrape, wash, disinfect and dry the calving stalls before you turn in the springers. Fill the stalls with ample straw bedding. Shavings and sawdust are poor choices for bedding material because disease-laden dust is drawn into the uterus during calving. Be aware that a cow can become trapped in a stall with uneven floors. She may get her feet uphill from her back and not be able to get up.

Watch Udders for Edema

When your cows are close to freshening, move them to the hospital area or parlor so you can examine their udder. Just before calving, springing cows have a lot of udder swelling. Look for heat, pain or unbalanced swelling in one quarter; these signs indicate mastitis. Monitor all dry cows regularly for evidence of udder edema. If it is common, you may need changes in your dry-cow ration to remedy the problem. If edema is occasional, first check potassium levels and the cationic ratio in the dry-cow ration (see Chapter 19). When potassium levels increase or the cationic ratio widens, udder edema always increases. Next, determine how much energy you're feeding. Mature dry cows aren't nearly as prone to edema from excessive dietary energy as heifers are. Frequently dairymen feed heifers extra energy, in an effort to overcome stunted growth. On high levels of corn silage, heifers will develop serious udder edema, so limit this type of feed for heifers in late pregnancy.

Another way to counter udder edema is to limit the dry cow's salt intake so she doesn't retain water. Use a salt block instead of loose salt. You may even resort to using an udder support on an especially valuable springing cow (Fig 4). Otherwise, excessive udder edema may ruin her suspensory ligament. Don't be afraid to encourage a cow with edema to get plenty of exercise.

Some individuals may need milking before freshening. Put them through the parlor every day and palpate them so you know when the udder begins to get tight and hard. Once you begin milking, continue for every milking. The incidence of milk fever in these cows is higher, so look for early signs. It's less of a problem to treat milk fever than it is to deal with a cow whose suspensory ligament has given way, or one who has developed mastitis. If you've milked before calving, the cow won't have quality colostrum available for the calf. This is the time to thaw out some frozen colostrum.

A different problem is the cow that leaks milk continually before she calves. She will be susceptible to mastitis because the teat sphincter is

Figure 4. If untreated, severe udder edema may stretch the suspensory ligament and cause the udder to become pendulous and prone to injury. If edema medications prove ineffective, an udder support, such as this one, can provide significant relief.

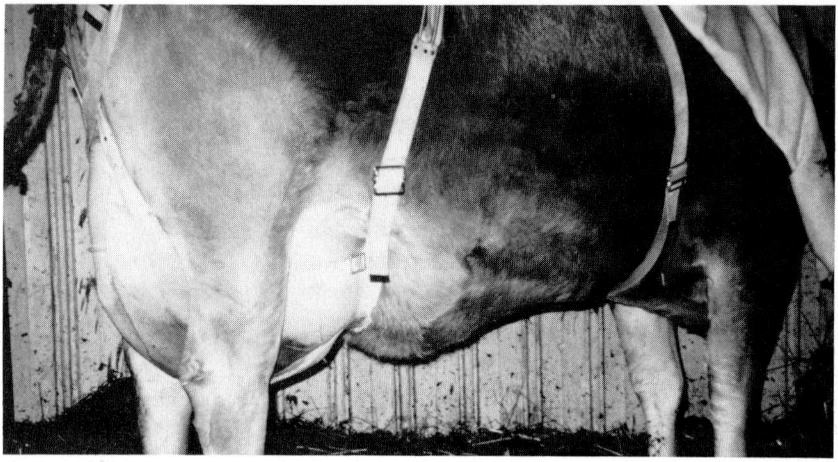

open. Some dairymen think barrier teat dips will stop the leak, but teat tape works better because it seals off the teat end more securely.

References

1. *J Dairy Sci*, 1984. p 642.
2. *J Dairy Sci*, 1984. p 1717.

7

Managing Freshening Cows

Hormones are the fuel that drives calving. In the week before freshening, the cow's estrogen levels increase, and progesterone levels decrease. There is also an increase in relaxin, which relaxes the pelvic ligaments. These hormonal changes make the cow look loose and sunken around the tail, and stretched around the pins.

Various hormonal changes also occur in the fetus. The fetus produces a high level of cortisol, which is the signal to the cow's calving mechanism. This causes the release of hormones that relax the cervix and begin uterine contractions. If a cow is carrying a dead fetus, she may very likely be unable to start the freshening process. And in some cases, fetuses with defective adrenal or pituitary glands cannot secrete the necessary cortisol to trigger calving.

Calving

Stage 1

The first stage of calving begins with uterine muscle activity. Contractions begin before you are aware that the cow is in labor. You will notice, first, that the vulva is swollen. Next you may see a slightly blood-tinged discharge on the cow's tail, which is the mucus that sealed the cervix during pregnancy. At this time, if you see any part of the placenta, call your veterinarian immediately. It probably indicates that the placenta is detaching prematurely, the calf is dead, or that there is premature delivery without proper uterine contractions to dilate the cervix.

As the first stage of calving progresses, the cow (especially a heifer) may become restless, nervous or even aggressive. A cow just about to calve frequently stays away from the rest of the herd. If the cattle are on a pasture lot, she may find a secluded place in which to calve. Soon, regular uterine contractions begin to occur about once every 30 minutes, gradually increasing in intensity and frequency. Eventually, the uterine contractions occur every 5-6 minutes and finally at a frequency of 3-4 every minute. As the contractions progress and intensify, the cervix dilates as the calf is pushed against it. Once the placenta protrudes through the dilated cervix, the second stage of calving has begun.

Stage 2

During the second stage of labor, uterine contractions are strong. The calf is propelled into the birth canal. The cow becomes very intense, almost frantic, in her efforts to push the calf out. As the feet and head extend into the birth canal, the placenta may rupture. The head and feet protrude from the vulva. This is a sign that delivery is imminent and you probably won't need to offer any assistance.

However, if you have doubts about how the calving is progressing, this is the time to examine the pelvic canal. A normal delivery in a mature cow takes about 2-4 hours, and in a heifer a little longer. Trust your instincts if you believe the calving isn't going well. Here are some solid reasons for assisting: the feet and legs are taking a long time to come out; the feet aren't together when they protrude from the birth canal; and the placenta is dangling from the vulva.

There are several steps to follow when you examine the pelvic canal. First, restrain the cow. It's a simple task to use a halter to tie the cow to a tractor or truck bumper if she is in the middle of the field. Even if it seems like a cow is easy to catch and hold, by the time you're ready to manipulate the calf inside her, she may be up and running. Getting her excited may cause a release of adrenalin, which will stop labor; so tie her.

Second, clean up properly. Tie the tail out of the way, preferably to the cow rather than to a solid object close by (or she may very well rip her tail off). It's much safer to tie her tail around her neck or around her front leg. You can also have an assistant hold the tail out of the way. Clean all of the vulva, the protruding head or feet of the calf, the cow's anus, and the immediate hindquarters of the cow (Fig 1). Use soapy water and cotton or paper towels. Then clean your hand and arm and put on a new disposable sleeve. Spread some clean lubricant over the sleeve. Now you're ready to put your hand into the cow's vagina and feel how the calf is presented.

Delivery Problems

Problems Associated with the Cow

Various problems may occur during delivery. When a cow goes down with milk fever before calving, she may not be able to deliver the calf because the uterine muscles are lacking calcium. They are one of the first muscle groups to be paralyzed in milk fever. Even when the calf is

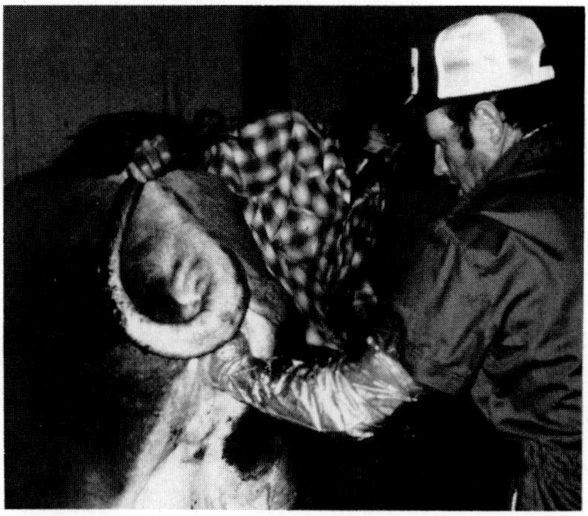

Figure 1. Cleanliness is of utmost importance when performing an obstetric exam on a cow. If you do obstetric exams on your cows, review your technique with your vet. Before the exam, wash the cow's vulva and pin bone area thoroughly with mild soapy water. Wear a disposable glove that is thoroughly lubricated with a nonirritating gel.

presented correctly, delivery can't occur on time because the cow can't push. Another potential problem with a freshening cow is a low selenium and vitamin E level. Deficiencies of this kind cause ineffective uterine contractions. If you have a number of slow deliveries, your herd may need more of these nutrients.

Fat cows may be in a weakened condition because of the fatty state of their liver. Excessive fat in the pelvic canal also can interfere with delivery. On the other hand, a poorly nourished cow may be so weak that she can't have strong uterine contractions. And one that has had pelvic fractures may now have a pelvis too small for the calf to pass. This is also true in poorly grown heifers. They frequently have a small pelvis, which means difficulty in delivery. Stunting early in life can't be "fixed" by proper feeding during pregnancy.

Occasionally a cow may develop a condition called "hydrops," more commonly called "dropsy," which causes excessive fluid to accumulate in the uterus. The calf may be of normal size but the cow, because of the many gallons of fluid, may appear to be carrying twins. Another type of hydrops, called hydrops amnion, results in a very thick, tenacious fluid in the uterus. In this case, the cow goes down and may die. Fortunately, neither type of hydrops is common. When it does occur, your vet can usually treat it.

Problems Associated with the Calf

Large Calf

The most common delivery problem is a calf that is too big, relative to the size of the cow. In moderate cases, you can pull the calf fairly

easily by using chains (not baler twine) on the feet. When you can't seem to pull a big calf with moderate traction, resist the temptation to get out the tractor. You can ruin a cow using this method. Call your vet instead. He'll make an incision along the side of the vulva toward the pin bone to allow more room in the area of the vulva. This is called an episiotomy.

Delivery by cesarean section may be necessary for a big, live calf. If the calf is dead, the calf may have begun to rot inside the cow, thereby increasing the risk of infection in the cow. Your vet will have to perform a "fetotomy." This means the vet must cut the calf into pieces within the uterus and deliver them, one at a time. It's tough to have lost the calf, but this procedure can save the cow.

". . . so as soon as I seen the old gal havin' trouble calving, I called you right away, Doc."

Abnormal Presentation

A second delivery problem is a calf presented backwards. This is called a posterior presentation (Fig 2). A cow can't be effective in delivery when her contractions are pushing on the head and shoulders of the calf, rather than on the rump and hind legs. If the calf is moving straight backwards, you'll need to provide some traction. Attach chains on the hind legs and pull as the cow pushes. Angle your pulling toward the ground at a 45-degree angle to the cow. Keep in mind that the calf's tail sometimes gets bent up over the back, and may cause more resistance.

"Wait! Hold everything! There's something wrong here!"

In other backward presentations, the calf is presented breech. This means that both hind legs are pointing forward in the cow with the rump and tail coming out first (Fig 3). It's common for a cow carrying twins to have the first one presented breech. She can't deliver either one because the first one isn't leading with its feet. When I see a cow trying to labor with only the calf's tail protruding from the vulva, I get an empty feeling in my stomach, especially if the cow has been pushing for any length of time. It's not hard to manipulate a breech calf into a better position if the cow has only been laboring a little while. But it's a totally different story when she has been pushing for 24 hours.

Calves presented frontward or backward have to be delivered feet first. You can usually tell how a calf is coming by looking at the feet. If the top of the hooves face up, it probably means the calf is presenting its front feet and head. This is called anterior presentation. If the feet are turned with the heels up, it likely means that the calf is coming backwards.

There are exceptions to both of these rules, such as when a calf is coming backwards and upside down as well, so that the back feet appear

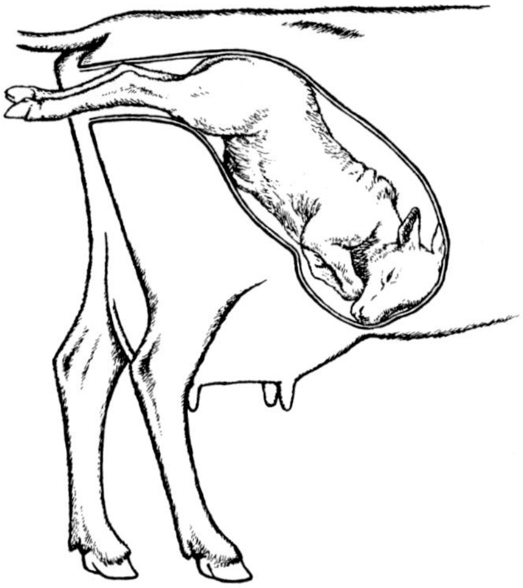

Figure 2. Most cows are unable to deliver without assistance when a calf is presented with its hind feet first. A calf in posterior presentation can be delivered with traction without changing the calf's posture. One person must pull the calf while a second person guides the hindquarters so that the rump and tail are not caught at the top of the birth canal. Traction should be applied at a 45-degree angle to the back of the cow.

like front feet in an anterior presentation. To clear up the mystery, do a vaginal examination so you can feel the rest of the calf's body. It helps to become familiar with the difference in shape between the front and hind legs. Be aware that the front legs have the knee joint between the foot and the elbow. The hind legs have no joint between the foot and hock joint.

Another presentation you'll see is the calf coming backward with one leg back (Fig 4). Calves also may be positioned backward and upside down.

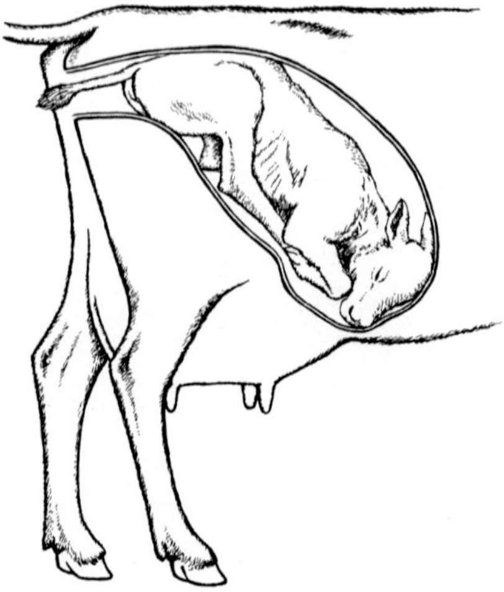

Figure 3. In a breech presentation, the calf is presented posteriorly with both hind legs pointing forward and only the tail in the birth canal. The calf's rear end must be pulled up into the canal for delivery. Breech presentations are very common with twins.

Figure 4. Occasionally, a calf faces backward with one leg in the birth canal and the other pointing forward. This type of delivery is made by pushing the calf forward so the folded leg can be drawn back into the canal.

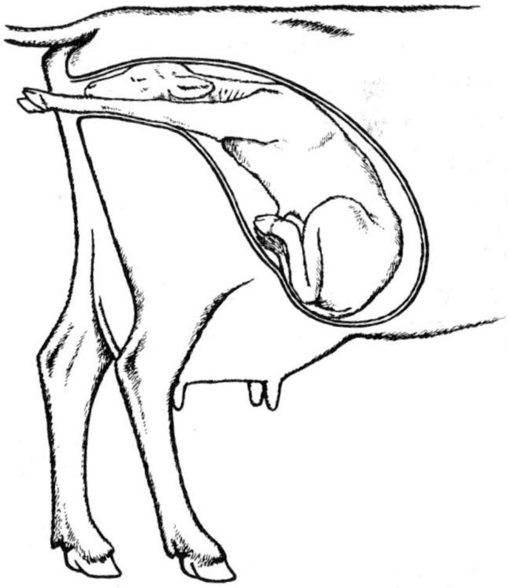

Figure 5. Normal presentation of a calf during delivery.

You're sometimes faced with positioning problems even when the calf is presented frontward. One front leg can be folded back, or both legs can be folded back (Fig 6). One front leg and the head can be back, or just the head can be tucked back. The important thing for you to remember is that the front legs and head must be delivered together (Fig 5). There is nothing more frustrating than to answer a dairyman's call for help and discover that he pulled on the calf's feet without first finding out by pelvic exam that the head was doubled back (Fig 7). With the head turned 180 degrees and the feet out, any vet would have an extremely difficult time manipulating the calf to deliver with head and front feet together. In these cases I'm often faced with doing a cesarean section to salvage a living calf or doing a fetotomy on a dead calf.

Figure 6. A calf in anterior presentation can become stuck if one or both legs are doubled back. This type of calf dystocia is usually corrected by pushing the head and the leg back inside the cow and drawing the legs forward. Once both legs and the head are presented together, most calves can be delivered successfully. If both front legs are folded back, many dairymen have difficulty pushing the calf's head back into the birth canal. To manipulate the head back, the calf's forehead must be in direct line with its back. When the forehead is held in line with the back, pushing with your palm on the calf's muzzle usually allows the head to be pushed back in. If the calf is dead, its head may need to be removed to reach the legs and extract the body.

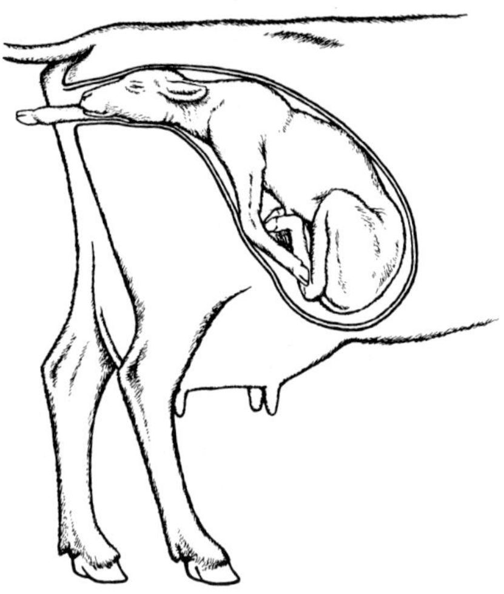

Correcting Abnormal Positions

Here are some directions for correcting the difficult malpresentations. Let's say the head is coming correctly, but both front feet are tucked back. Reach past the head and pull up each leg with obstetrical chains so that they can match the progress of the head. With a small

Figure 7. One of the most common types of dystocias is an anterior presentation, with the head doubled back. In some cases there is too little room in the birth canal, but too often a dairyman has been pulling on the feet without first checking to see if the head is in correct presentation. When this occurs, it becomes very difficult to correct the positioning and deliver the calf successfully.

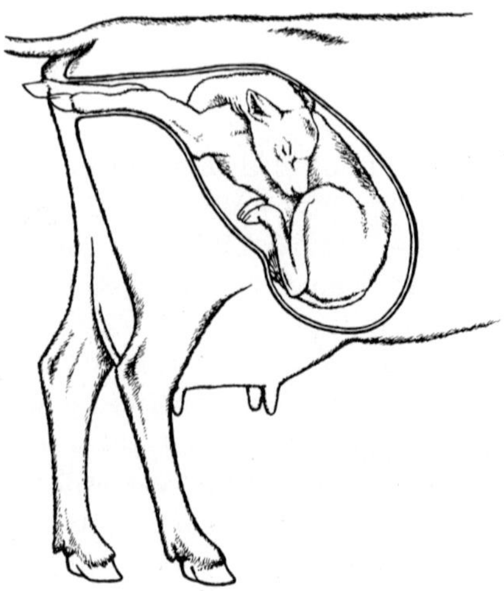

heifer or a big-headed calf, there may not be enough room to reach past the head to the legs. It's also hard to manipulate the calf if the head is already delivered past the vulva. In these cases, you must push the head all the way back in the birth canal so you can reach the legs and attach the chains. Push the calf's nose upward so the forehead is parallel to the cow's backbone. Then put the palm of your hand on the muzzle of the calf, line up its forehead with its spine, and push the head straight back in. In most cases this takes a lot of strength, especially if the cow is pushing against you. If the calf is dead, sometimes it is easier to cut the head off to retract the legs.

Calves presented upside down are difficult to turn over, but you want the chin resting on the feet. The best way to flip the calf is to pull on the calf as you twist the body to turn it over.

When a cow is laboring and you find 4 feet coming out, you or your vet need to do some detective work before you do any pulling or manipulating. In most cases, there are twins in the cow. Rarely does a single calf present all 4 legs and no head. The converse situation to a calf's presenting 4 legs is for it to present the middle of its back first. You can deliver the calf in either the anterior or posterior position, depending on which end you can reach most easily.

When a cow is delivering twins, you have to know some tricks. If the first calf is coming backward and breech, you can usually push the calf's rump back in and straighten its legs out. As I mentioned before, this only works when you work with the cow early in labor. Sometimes 2 calves may be presented properly, but both are coming at the same time. You'll need to feel which set of legs goes with which twin, then deliver one calf while you repel the other.

In cases of hiplock (also called stifle lock), you have a special challenge. In these cases the calf's stifles are too large for the cow's pelvis, and they get hung up on the brim of the birth canal (Fig 8). It's quite a common problem, and one you may want to turn over to your veterinarian. If you try to tackle it, twist the calf in one direction or the other at the same time you're applying traction. Pull downward at a 45-degree angle to the cow's rear end. You may need to pull first to one side and then the other.

Occasionally the only way to relieve this condition is by brute force, but you risk paralyzing the cow. Another option for a small heifer is to split her pelvis—a task for your vet. He'll make a small incision below the birth canal in the skin below the vulva. Then he drives a chisel into the pelvic bones so that the pelvis is split. This usually expands the birth canal an inch, which may be enough to allow the calf to pass through. These heifers often recover without problems, but it's necessary to hobble their hind legs temporarily until they heal. Keep them off slippery

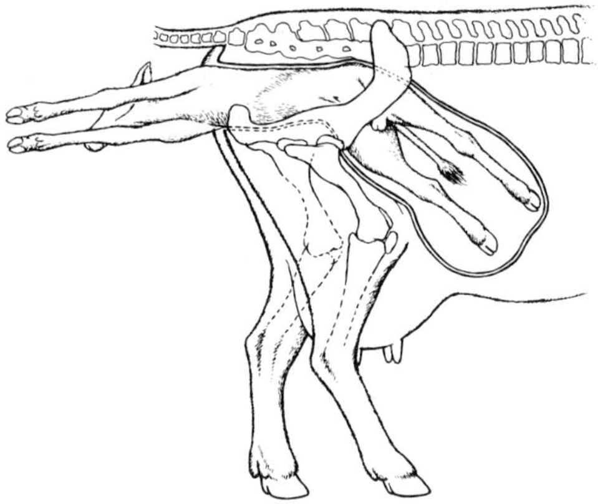

Figure 8. Perhaps the most frustrating type of delivery is a "hiplocked" calf. The calf's stifles catch against the cow's bony pelvis. The calf is alive when it becomes stuck, but usually dies during delivery. The calf must be twisted from side to side to be freed. Sometimes changing the angle of traction helps. Rarely will your vet need to deliver the calf a piece at a time.

concrete until they regain full use of their hind legs. The hormone relaxin may soon be available to cause greater expansion of the pelvis.

Fetotomy is also an option in a case of hiplock. The vet may have to cut the calf in half, push the pelvic portion back into the cow, then deliver the front portion. He'll then split the remaining portion with a fetotomy wire before delivering it.

Fetal Abnormalities

No discussion of calving difficulties would be complete without mentioning fetal monsters. These are calves that have developed abnormally while in the uterus. They may have 2 heads, 5 legs, 2 bodies or any combination of problems that lead to a deformity. Some calves' joints are frozen in one position so you can't straighten the legs for delivery. Another extreme example of a fetal monster is a calf that has literally grown inside out. The legs, head and skin are directed inward on the calf and the intestinal tract and sometimes the heart and lungs are outside of the calf. As you might expect, a cow can rarely deliver a calf like this without assistance.

Cesarean Section

With attentive care, cows that labor no longer than 4 hours usually will deliver live calves. The ones you can't deliver successfully with traction are candidates for delivery by cesarean section (C-section), a procedure for delivering the calf through the cow's side. I do this surgery when a cow has a live calf inside and we want a live calf. You put the dam at some risk because this is major abdominal surgery. But the risk is offset somewhat by the good odds of getting a live calf.

Your vet can perform this procedure with the cow either standing or lying down on her side. Provide a clean room or stall for the surgery. Clip the cow's flank or midline free of hair, and the vet will prepare the incision site. Then he'll cut through the body wall and uterine wall, and deliver the calf. Then he closes the uterine wall and the tissue layers of the body wall with sutures. When he does this surgery under sterile conditions, the cow will likely recover without complications. Cesarean sections may not be very successful when the calf inside is dead, especially if it has been dead for some time.

Prolonged Labor

Unfortunately, every dairyman sees his share of dead calves, and sometimes these dead calves cause delivery problems. A cow can deliver a dead calf without problems if she hasn't labored too long. But when she pushes for more than 4 hours after the placenta ruptures and the amniotic fluid passes, the odds are high that the calf will die from the delay. This is because once the placenta separates from the uterine wall, the calf loses its source of oxygen. The cow's uterus quickly becomes fatigued as well. Then the uterine muscles clamp down on the calf. This reaction is called Bandl's rings. The uterine muscles are so taut that you can tear the uterus just by putting pressure on it in an effort to stretch the birth canal. This makes it difficult to pull a calf from a cow. I end up doing a lot of manipulation on the calf, while being gentle to the cow's uterine wall.

Fetotomy

When the calf is dead and traction won't help the delivery, fetotomy is another option. The vet must cut the calf into pieces and deliver them one at a time. This procedure poses less of a risk to the dam than a cesarean section, she has a better chance of recovering fully, and she's more likely to get pregnant on the next breeding than a cow that has had surgery. The cost for the 2 procedures may be about the same.

The vet does a fetotomy to reduce the size of the calf so it can come through the birth canal of the cow. It can be a gruesome task but necessary to save the cow.

Slaughter

The last option for dealing with difficult deliveries is to slaughter the cow to avoid putting her through the misery of some of these procedures. Cows suffering this badly have often been neglected, and I hope this option isn't one you'll often have to consider.

Stage 3

In the last stage of labor, the placenta passes out of the uterus and the uterus begins to regain its original shape for normal fertility and subsequent breeding. The placenta should be passed 1-2 hours after delivery. If it takes 8-12 hours, I consider it a retained placenta.

A cow that has had a difficult delivery usually doesn't pass her placenta readily because the uterus becomes edematous from manipulation. Cows that calve early, or have twins or systemic disease often have retained placentas, also. Various drugs aid passage of the placenta. While experts believe oxytocin doesn't stimulate placenta passage, it does stimulate uterine immunity. Ergonovine, a derivative of ergot, causes uterine contractions, but there is no current research to confirm its positive effect. Most recently, a drug called PGF_{2a} has been recommended for treatment of retained placenta.

Once the placenta passes, the uterus gradually shrinks in size. By the seventh or eighth day after calving, it's almost totally empty of fluids. By the tenth day after calving, you will notice increased passage of tissue debris and fluid. This discharge may appear very bloody, but it's normal. This is how the uterus naturally cleans itself. You don't need to infuse the cow. Infusing at this time may actually weaken the uterine immune system.

By 10-14 days after calving, a follicle develops on an ovary and the cow has her first ovulation. In most cases, she doesn't show signs of heat at this time. The uterine horns shrink greatly, which helps expel fluid from the uterus. Once the first ovulation has occurred and the fluid is passed, you can expect little discharge from the uterus. Most dairymen and vets consider uterine therapy if they observe a copious and foul discharge by the 14th day.

At day 22, the uterine horn diameter is smaller than that of the cervix. Horns that are bigger than the cervix by this time need treatment. By day 30 the uterus feels like one in a cow with normal fertility.

Watch for Prolapsed Uterus

A condition that sometimes occurs in the third stage of calving is a prolapsed uterus. With uterine proplapse, the cow passes her uterus inside out through the vagina. This usually happens right after freshening, though it may occur up to 24-36 hours after calving. The affected cow goes into shock. Sometimes the prolapsed uterus becomes lacerated and bleeds if she walks around. After prolapse, it usually swells because of impaired blood flow. The longer it is inside out, the greater the swelling. Occasionally the cow strains enough that the bladder or some of the intestinal tract extends into the everted uterus. This makes reduction difficult.

There are several reasons why a uterus prolapses. It happens most frequently in a cow with milk fever. She is constipated and does a lot of straining while she is down. A cow that has had a difficult birth may also strain, and this muscular activity can push the uterus up into the birth canal. Calving causes the ligaments in the pelvic area to relax, so the uterus can easily slip through the vagina.

A cow may be prone to prolapse if she's been paralyzed because she delivered a large calf. You find her down on her side, straining, and unable to get up. A large, heavy cow that's down because of fat cow syndrome also may suffer uterine prolapse. Try to avoid letting a freshening cow get trapped on an uneven floor where her feet are higher than her back, or her rear end is downhill from the rest of her body. Cows can get trapped under freestalls or gates, too (Fig 9). Heifers that have eaten moldy feed may prolapse their vagina, and sometimes their uterus.

Treatment

When you've identified a prolapse, call your veterinarian immediately. This is an emergency. If the cow is standing up with a prolapsed uterus, it's wise to move her to a clean place, and restrain her so she doesn't lacerate her uterus. Bed down the cow on clean straw. If she is

Figure 9. This cow injured her hip in a freestall while calving and died of pneumonia despite intensive nursing care and physical therapy by lifting. The loss of this cow could have been avoided if she had been in a calving lot or maternity stall when she calved.

Figure 10. A board or small table placed over the backward extended legs of a down cow works well to support a prolapsed uterus at the level of the vulva. If the cow is on her feet, 2 assistants are needed to hold the board and uterus at the proper level for reduction.

down, she should have her hind feet straight out behind her. This reduces the straining. Put ropes on the hind feet and pull the legs backward. This makes the pelvis tilt downhill, and the uterus will more readily fall into position. Standing cows have this advantage naturally.

Clean the uterus while you wait for the vet. You can apply sugar or alum to help relieve its swelling so it'll be easier to insert. Have clean, warm water ready for the vet. And if you can round up another helper, the 3 of you can get the job done more easily.

With standing cows, I use a clean 16-inch-wide board to support the uterus at the level of the vaginal floor so I can push the uterus back in (Fig 10). With down cows that are positioned properly, I put the board across their extended hocks and rest the uterus on it. This is also a good place to get the uterus cleaned up and ready to put back in place. Your vet will replace the uterus by pushing the mass back into the vagina, gently putting in one side at a time. He'll use his fist rather than his fingers to avoid punching a hole in the uterus. This is often very difficult.

When I talk about prolapses, I'm always reminded of a colleague of mine. He was faced with an especially serious case; the cow was in guarded condition and the swelling was severe. He found it hard to differentiate the opening of the vagina from the rectum in this case. After several minutes' work, the cow died. At that moment, my colleague discovered he had been trying to stuff the uterus back into the rectum. It might have been interesting to see the results if the cow had lived!

Some cows die just as the uterus is replaced. It's not because the vet made a mistake, but because restored blood flow carries toxic, stagnant or clotted blood to the cow's brain and heart. The vet keeps the re-inserted uterus in place with prolapse pins, Bruner sutures or other appropriate means. Otherwise, the cow may strain and push the uterus right back out again. The therapy for the cow recovering from a prolapsed uterus is solely dependent on how sick she is. A cow in shock is likely to need intravenous fluids, cortisone and other supportive medication, including antibiotics. In the past, vets often used intravenous tetracycline, but today a better choice is large doses of penicillin G. In any case, keep the recovering cow from running with the rest of the herd for a while. She deserves a little time to herself and special care to see that she gets on feed and recovers quickly.

"Doc, would it help if we smeared it all over with some of my Preparation-H?"

Watch for Prolonged Gestation

A relatively rare condition in dairy cattle that relates to advanced pregnancy and calving is called prolonged gestation. It stems from a developmental defect in the calf. The cow will be pregnant for 10-11 months and has difficulty calving because the calf continues to grow and the cow continues to get bigger. The calf may weigh 300 lb at birth, so it's not surprising that the cow has delivery problems. The cause of prolonged gestation is an adrenal or pituitary gland defect in the calf.

The key to treating prolonged gestation is to trigger the mechanism that initiates calving. We know that the cortisol level must rise to initiate labor. So we induce calving with an injectible cortisone, such as dexamethasone. The vet will often have to do a cesarean section or fetotomy. In either case, the calf won't survive once born. Unfortunately, there's no known way to prevent this condition.

Lactation without Calving

Artificially inducing lactation in a nonpregnant cow makes a cow freshen without having a calf. This allows cows with serious reproductive problems to become productive. It may help the valuable cow that needs more treatment to get in shape for breeding. Milk production reduces the extra fat on her back and conditions her for breeding.

There are also some disadvantages. The procedure requires giving a number of estrogen injections over a 3-week period. Milk production may not be as high as with a natural freshening. Also, the FDA has not formally approved the use of estrogen for this procedure. Estrogen has been shown to be a cancer-causing agent, and the FDA restricts its use on lactating cattle. Until residue problems are resolved, the procedure won't be available to most dairymen. The DHIA may not accept production records on cows that have begun milking without having calved.

Caring for the Newborn Calf

Once you've got a live, healthy calf on the ground, here's how to proceed. Clear the calf's airway of membranes and mucus with your hand and a towel, and then stimulate the calf to breathe. All you have to do is put a straw or pencil into its nostrils deep enough so that it coughs and sneezes. If this isn't enough, you can slap the calf on the chest. Usually the heart has been beating adequately throughout delivery. When there's more mucus in the respiratory tract than you can wipe away, you may have to pick up the calf by the hind legs and swing it back and forth several times. In more severe situations, you can give oxygen using a little tank like the ones used for people with asthma and emphysema. Your vet may need to give respiratory stimulants, such as doxapram, along with the oxygen. This calf often will have acidosis. If he is responding poorly, this is likely an indication that your vet may also need to start an IV to correct the acidosis.

Once the calf's breathing has stabilized, inspect the umbilical cord (navel). If it is bleeding, clamp it off with forceps or vise grips or tie a thread around the cord tightly. Next, dip the umbilical cord in a 2% iodine solution. Do this several more times during the first couple of days until it dries up.

The calf needs colostrum within 6 hours of birth. This first milk from the cow contains all of the antibodies and immunity the calf needs for a healthy early life. Most large-breed calves need about 3-4 quarts. However, many calves won't take this much colostrum at one feeding, so give several smaller feedings in the first 6 hours. For calves that won't nurse a bottle, you must use an esophageal feeder (Figs 11, 12). Every dairyman should own one because you can use it without fearing you'll put a tube into its lungs instead of its stomach on an upright or standing calf.

The calf's body weight can be used as a guide to the amount of colostrum required. A calf should get at least 5-8% of its body weight in colostrum during the first 6 hours of life. For example, a 100-lb calf should receive 5-8 lb of colostrum.

There is variability in the quality of colostrum. A cow that has leaked milk for a week or so before she calves may have poor-quality colostrum. Because heifers haven't been exposed to the same diseases as cows, they have lower-quality colostrum. To give optimum immunity, you may

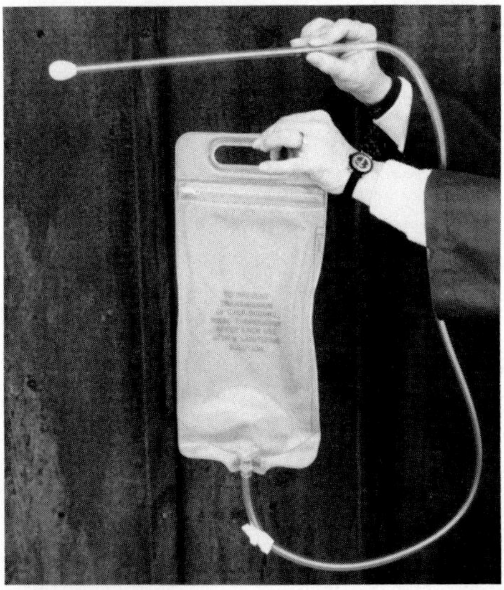

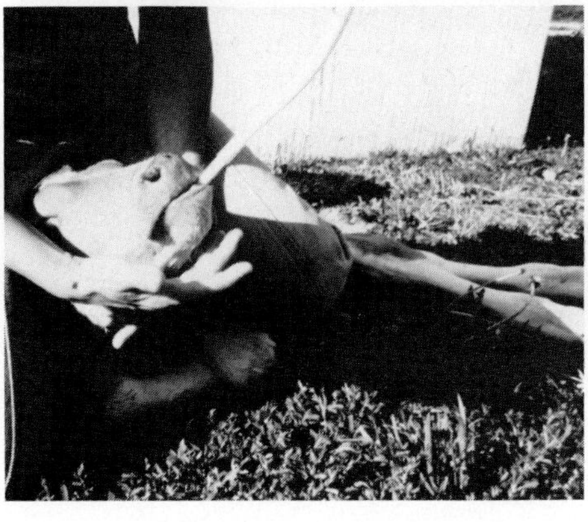

Figures 11 and 12. Proper use of an esophageal feeder can ensure that every newborn calf gets the proper amount of colostrum. Calves that are ill and refuse to nurse can be given needed nutrients with the feeder. When using the feeder, hold the calf in an upright position so the rigid tip of the feeder slides into the esophagus and not into the windpipe.

decide to give high-quality colostrum from a cow to heifers' calves. It's easy enough to collect and freeze cows' colostrum for use at a later date. It can be stored frozen for a year or more. I suggest you thaw each package in a microwave oven on the low setting.

One last task you may need to perform after a calf is born is treating a fracture. Sometimes when you use traction to aid a delivery, a calf's leg may be broken. Your vet will need to fit the calf with a splint and keep it in place for 4-5 weeks to allow healing. Keep the calf in a dry place so that the splint stays in good condition. When the calf outgrows the first splint in 2-3 weeks, change to a bigger one. In rare cases, this calf may have suffered enough damage from the traction to destroy blood circulation to its foot. The hoof will slough, and there's nothing that can be done to repair the damage.

8

Raising Heifers and Purchasing Replacements

Purchase, Lease or Raise?

In this new era of dairying, many of the old rules may no longer apply. You must regularly evaluate new options that in the past you wouldn't have considered. One of these is whether to raise or purchase replacements. There are a number of factors to consider.

If you elect to purchase replacements, you can regulate milk production to when extra production time is needed. There are often bargains on heifers with newer genetics. This is especially true with embryo-transfer heifers. If you are in a milk quota base marketing program, replacements can balance your production for optimum results.

Now is the time for a critical appraisal: Are you doing a good job of raising replacements? If not, consider having someone else raise them, and purchase the heifers when they are ready to freshen. Sometimes it is advantageous to lease heifers. This may be a better alternative than financing purchases. Another element to consider is the cost of raising replacements versus purchasing or leasing them. Do you have room in your facilities to raise heifers?

A number of disadvantages must be considered when evaluating the choices. Leased cows often have a high incidence of health problems. You must evaluate these new animals carefully before bringing them into the herd. Generally, this should include the SCC as well as reproductive status. Your vet can help you with a decision. If new animals aren't checked thoroughly, you may introduce a disease into the herd with your additions. Bringing new heifers or cows into the herd also disrupts the social order. If the new animals are not introduced carefully into the herd, you may lose some of that anticipated increase in production.

All new animals are stressed by placing them in a new environment, especially if they arrive when they are about to freshen. Consequently, they are prone to retained placenta and metabolic disorders. Milk production in cows purchased in midlactation is likely to drop drastically when these cows are brought into your herd.

Here are some guidelines for purchasing or leasing replacements.

- If you have any doubts about the cows you're purchasing, make the purchase subject to veterinary examination. Don't rely on the seller's vet. While he's probably ethical, he'll have the seller's interests at heart.
- Evaluate the health of replacement cows. Palpate the udder and check the SCC. Also check the SCC of the herd. Inquire about the health of the herd. What is the vaccination program?
- Request an official health certificate signed by the vet who performs the exam or certifies the health status. These papers should include the results of all tests, such as TB and brucellosis. If you aren't familiar with your state's regulations, ask your vet.
- Do not accept delivery of cows without verifying their health papers. When replacements arrive, have your vet examine them. Check fertility or pregnancy status. Check for pinkeye and other potential infections. If there is a problem, the seller should be contacted immediately rather than after the problem progresses.
- Replacement animals should be quarantined, if possible, for 3-4 weeks. Seek your vet's advice for your situation.
- When selecting replacement heifers or cows, be very careful. Purchasing replacements is often no different from buying a used car. Without careful selection you may purchase a "pig in a poke."

Raising Heifers

Replacement heifers are essential for improvement of your herd. Most dairymen assume that they ought to be the one to raise replacements, but in fact, many do a poor job. If you uncover the hidden costs in raising heifers, you'll probably be faced with the choice of buying replacements or knuckling down to improve your own program. Putting top-of-the-line replacement heifers into your herd on a regular basis is a sure way to increase herd production and profits. The profit potential of a heifer is a direct reflection of the skill you used in raising her.

From Birth to Weaning

First Feeding

No matter what feeding program you use, colostrum is essential to the newborn calf. This first milk from the cow contains antibodies that provide the calf with passive immunity to disease (see Chapter 7). By hand feeding, you make certain the calf takes in as much colostrum as it needs. Left to fend for itself, the calf may not get the "full dose" of antibodies because it sucks poorly, its dam won't stand still for feeding, or the shape of her udder makes it difficult for the calf to nurse. Feed the colostrum within 6 hours of birth, because the calf's digestive tract gradually loses the ability to absorb the antibodies. Figure on delivering about 5-8% of the calf's body weight in colostrum during this time. Recent data suggest that a calf's colostrum absorption will be poor if it was in dystocia for more than 1 1/2 hours. The reason for poor absorption of antibody is that such calves often have acidosis (see Chapter 5). You can improve this calf's chances of surviving by having your vet correct the acidosis with an intravenous infusion.

Most colostral antibodies last from several days up to 6 months in the calf. Another kind of immunity a calf can benefit from is active immunity. This immunity develops after a calf is exposed to infectious microorganisms and then produces antibodies against those organisms. Active immunity isn't immediately available at birth like passive immunity; it takes about 2 weeks to develop after exposure to an organism. It can last from a year to a lifetime.

Without passive immunity, a calf can develop a number of problems. It may develop septicemia, an overwhelming infection that produces toxins that can kill a calf in a matter of hours. This is often an *E coli* septicemia. Note that it may be a different syndrome from the *E coli* scours that calves get. A calf can acquire *E coli* septicemia by ingesting the organisms, which then multiply in the bloodstream. Or if a calf doesn't receive enough antibodies from colostrum, it may develop arthritis or meningitis. The calf appears lame or deranged and can't stand up. Most affected calves die.

To avoid these losses, feed high-quality colostrum immediately. One way to assess the quality of colostrum is to use a hydrometer. This instrument measures the specific gravity of liquids. If your sample has a reading of less than 1.050, colostrum quality is suspect. A new device, called a colostrometer, has a chart to confirm colostrum quality.

Heifer colostrum is usually lower in quality than cow colostrum. If you want to use heifer colostrum, mix good-quality colostrum from cows with the colostrum from the heifer. Keep in mind that the antibody level in colostrum drops rapidly after freshening, so get that very first milk to calves for their first meal.

Once a calf has had its first meal, either by nipple bottle or esophageal feeder, you may want to feed the calf out of a bucket. This helps eliminate the sucking response so that your calves won't suck on each other. If you feed with nipple bottles, house the calves separately.

Housing

I recommend small, portable calf hutches — structures measuring 3 x 5 ft or 4 x 6 ft. Each calf has its individual house and cannot associate with neighboring calves. Space them several feet apart so each calf lives in its own environment. Some dairymen build their own hutches of wood, plywood or sheet metal, with fiberglas roofing. Or you can buy pre-made fiberglas hutches. In comparison studies of wood versus fiberglas construction, wood structures were cheaper, but they didn't last as long as the fiberglas ones.

Locate the hutches away from the herd on a bed of large, crushed aggregate rock so that water can drain away. Bed the calves in straw. Choose a site close enough to a building so there's a natural windbreak. In areas of the country where the temperatures routinely get down below -20 F, it helps to set calf hutches inside a machinery shed or close to other natural windbreaks during the worst of the winter weather. Another alternative to protect the calves from wind is to stack bales of straw around them.

A second type of housing to consider is grouping the calves in a barn with the cow herd. This used to be the popular way to raise calves; dairymen put calves in nooks not used for anything else. Unfortunately, ventilation in these barns is usually poor, and calves suffer respiratory ailments. They readily pass any illness to others in the group, and a disease to which the cows are immune may attack the calves. Often, the cows pass pneumonia and diarrhea infections to the calves.

Yet another type of housing available is the modular calf building. While it provides an environment away from the cow herd and is convenient for feeding if it's outfitted with hot water and sinks, it has disadvantages. When one calf gets sick, the infection can pass easily to others nearby. Then you must disinfect the building before you bring in another batch of calves. One other suggestion is to provide calves in modular buildings with individual feed and water buckets that you clean on a regular basis. This eliminates the possibility of spreading an infection throughout the facility by feeding more than one animal from a bucket. Most of these buildings don't have good ventilation, either. The fresh air that comes in along the ridge isn't tempered before it heads down to the calf pens. So dairymen sometimes slow down the variable-speed fans to reduce this draft, and then there's not enough mixing of the air to produce adequate ventilation. Odors and gases tend to build up from the manure pit underneath the floor.

An On-Farm Example

Charles used to sell modular calf buildings for a local dealer. He had several distributed throughout the area and was proud of the top-notch job he was doing. He put one of these buildings on his own farm and bought bull calves from the stockyards to raise. Before long, he experienced lots of problems raising calves. Within 10 days, every calf he put in broke with severe diarrhea that killed about 50% of them. He was very frustrated and called me in to take a look.

I found that Charles' calves had BVD. Every time he put a new calf in, he exposed it to the disease. In the past, many veterinarians had recommended an exotic antibiotic, immune serum and BVD vaccination to clear up the problem. In most instances, this recommendation failed. I suggested that Charles break the cycle—stop putting in new calves and clean out the building. But as soon as one calf came down with BVD, he'd have the same vicious cycle again. So Charles opted for hutches.

Not right away, of course. He reasoned he ought to use a modular building if he was going to continue to sell them to other dairymen! I was rather frustrated with his attitude, especially when I thought about all the calf buildings that he'd sold all over the country. But I was heartened the next time I drove by. He had calves cozied into calf hutches out in his barn lot, and he'd hung a sign indicating he was now selling hutches!

Some of you will stick with modular calf buildings or permanent buildings in spite of my story about Charles. So to increase your odds of raising healthy calves there, divide your barn into rooms and design the stalls for an all-in, all-out rotation. Avoid continuously adding and weaning calves. Instead, fill a room and treat the calves as one group. When they reach weaning age, empty the room and disinfect it. You can cause calves to get diarrhea if you leave them inside while you wash down adjacent pens with the high-pressure washer. High humidity invites rapid growth of pathogenic *E coli*.

Ventilation

Attention to proper ventilation helps, too. In many calf buildings, high humidity causes moisture loaded with microorganisms to collect on the animal's haircoat. The hair mats and loses its insulating ability.

As a result, a calf is easily chilled, stressed and susceptible to diarrhea and pneumonia, especially if it didn't get a "full dose" of antibodies from colostrum. Air flow is equally important. Direct air away from the young calves, and from the calves to the cows (if they're housed under the same roof), not the other way around.

Neonatal calves are small animals and they don't yet have a functional rumen. Consequently, they radiate little heat. To lower stress in the barn in the winter, keep the temperature at 50-55 F. You may need to reduce air flow or add supplemental heat. This is especially true for calves raised in elevated crates without bedding. When you reduce air flow by turning fans on and off, the concentration of microorganisms in the air increases when the fans aren't running. Temperatures rise, too. When you cycle the large-capacity fans back on again, they suddenly pull cold air in and drop the temperatures. In effect, the barn fluctuates from being warm and more contaminated with "bugs" and fumes, to being fresher but suddenly colder. Calves are continually stressed by these conditions.

Calves will have fewer health problems when air changes are continuous at the rate of about 4 air changes per hour. It's important to install a duct from the fan to the floor so that the warmer air near the ceiling isn't exhausted and the colder air near the floor is removed. To maintain an even air flow without drafts, install a 1-inch (up to 4-inch) adjustable slot inlet at ceiling level in these rooms. It's best if you can pull room air from an attic or hay mow space rather than directly from outside as in the modular calf buildings. This tempers the air and eliminates cold shock.

For summer ventilation, provide fans controlled by thermostats so there are as many as 30-60 air changes per hour. A popular alternative in the Midwest is to use variable-speed fans. They're easy to adjust, but they're inefficient. Most cannot maintain a recommended static pressure of 1/8 inch at low speeds. This means that a 15-mile-per-hour wind blowing against the variable-speed fan may actually force exhaust air back into the building.

Cold Weather Feeding

Besides being more work for the calf feeder, hutches present another challenge. In periods of prolonged cold weather, calves raised on milk replacer may develop diarrhea and subsequently collapse in as little as 12-36 hours. This is because calves in the hutches don't have supplemental heat, so they must eat more to keep warm. In fact, these hutch calves need nearly twice as many calories in cold weather as calves in a warm building. It makes sense then to feed them nearly twice as much. Ignore this fact, and your calf soon develops a negative energy balance.

It burns reserve fat and body tissue to keep warm and eventually develops low blood sugar because it has burned all its energy stores. The first sign often is diarrhea in extremely cold weather.

The best way to treat this condition is to give intravenous glucose. Most calves will respond immediately, but, unfortunately, many affected calves die later. This is because most have depleted their body stores of energy. They use up fat and then start burning muscle tissue. Eventually they can't get up.

A rule of thumb for winter feeding in calf hutches in northern climates is to use whole milk and gradually double the amount you find adequate for summer feeding.

Milk Replacer

Regardless of the kind of housing you use, there are good rules of thumb for feeding calves. After you've administered the initial 5-8% of their body weight in colostrum, feed them a pint of whole milk daily for every 10 lb of body weight. Calves will fare well nutritionally if you feed them *once* daily, but without that extra feeding, you miss another chance to notice developing health problems. This is also true if you use automatic feeders.

In addition to milk, young calves need fresh water in front of them at all times. They'll drink larger quantities when they have diarrhea. Though there's a controversy about feeding mastitic milk to calves, recent reports indicate that it won't cause problems provided the calves cannot nurse each other after they drink it. But keep in mind that antibiotic residues in milk can cause residues in the calves or can upset fermentation in a sour colostrum program.

Milk replacer may be your choice of feed instead of whole milk. Avoid bargain brands because they often contain high levels of fiber from oat flour and distilled dry solubles. A calf cannot digest these ingredients until about 3 weeks of age. Replacer constituents should be all milk or milk by-products and contain about 20% crude protein, 15% fat (for warm climates) to 20% fat (for colder climates), and less than 0.5% crude fiber.

If a milk replacer is about 90% dry matter and you want to figure out how many parts of water to add, divide 90% by 12 (whole milk is 12% solids). This results in a liquid similar to whole milk. In this example, you need about 7.5 parts water to 1 part milk replacer.

The advantage of feeding milk replacer is primarily lower cost and the convenience of mixing up exactly what you need for the calves at any time. You can also avoid the diseases carried in raw whole milk. One disadvantage is that replacer is often not as wholesome for the calf as is whole milk. And you'll have a tougher time meeting the high energy requirements of certain calves with the replacer. Dairymen who find their calves don't thrive on milk replacer turn to other alternatives.

- Soured colostrum works well for some producers. You can ferment it without changing the feed value but you destroy antibodies present in the fresh product. Here's how to prepare and feed soured colostrum: Store excess colostrum in a 20-gallon garbage can or other suitable container with a lid to keep out flies and dirt.
- Begin to feed from a can after 3 days of fermentation. The fermentation process doesn't have to be completed.
- Use only 4-5 days of colostrum in each batch and stir well with each addition. Don't mix fresh colostrum with a supply that's already fermented.
- Keep 3 containers going: 1 container fermented and ready to use; 1 container in the process of fermentation; and 1 container that you're filling. Store all 3 at temperatures below 75 F. Warmer temperatures cause the protein to break down rapidly, lowering quality and palatability.
- Toss out any soured colostrum you've kept for more than one month.
- Always stir the liquid thoroughly before each feeding. Discard the whole batch if you see mold in it.
- Feed 2 parts of colostrum diluted with 1 part of warm water if most of the colostrum is obtained within 2 days of calving. If the cow is fresh more than 2 days, dilute 3 parts of colostrum with 1 part of water. Then feed 1 pt of sour colostrum/10 lb of body weight. Sometimes you can add 1 ounce of baking soda per gallon just before feeding to improve palatability.

Feeding soured colostrum is an economical method of raising young calves, but it requires some effort to monitor the colostrum fermentation process. You also may need to encourage calves that don't seem to like the taste of it.

Pickled colostrum is another choice. You preserve it by using an organic acid, such as vinegar, to lower the pH to about 4.6 and slow down decay. Use 1/2 cup vinegar for 4 1/4 gallons of colostrum. Another choice is 1% propionic acid. Add 1 cup of propionic acid to 6 gallons of colostrum. Collect fresh colostrum after each milking, add acid and stir. When you see the combination clabbered, stir it into colostrum previously preserved. The main advantage of pickled colostrum over sour colostrum is that it keeps better at temperatures above 75 F. The feed value is exactly the same. You may find that calves don't like the taste of pickled colostrum as well, but again, you can neutralize some of the acidity with baking soda just before each feeding.

Pathogenic *E coli* bacteria may be present in some fermentation or pickled colostrum in numbers that could hurt the newborn calf. You suppress bacterial growth in colostrum for a short period by fermentation or adding an organic acid. But ultimately, any preserved colostrum will develop bacterial growth at high temperatures.

Producers say the greatest benefit from feeding either form of colostrum is that it is partially digested before it gets to the calf. You use colostrum that would otherwise be discarded, and it's relatively easy to manage. With pickled colostrum, you don't need as many storage containers as for the sour product; once it's been pickled, you can add it to current supplies.

Here are some price comparisons: A calf will drink about 250 lb of milk from birth to 20 days of age. At $11.15/cwt, the cost of feeding whole milk comes to $33.45. About 30 lb of milk replacer would be needed to feed a calf for about 28 days. At $.50/lb, the cost would be $15.01. This is roughly half of the cost of feeding whole milk. If you used sour or pickled colostrum, you'd need about 163 lb to feed the calf. The primary costs here are labor and propionic acid. Counting the acid only at a penny a pound, you come up with a cost of $1.63 for the first 28 days. You'll have to figure in your labor cost.

Vaccinations

Calves may be exposed to a number of harmful microorganisms. Therefore, calves must be vaccinated to prevent infection. A serious infection to avoid is caused by *E coli* bacteria.

There are 2 steps to diagnosing *E coli* scours. First, you note the common clinical signs: diarrhea and depression. Your vet then follows up with a culture of the stool for *E coli* if your calves often have diarrhea. Keep in mind that *E coli* is readily found in most calves' digestive tracts, so lab tests generally detect the bacteria. The second step to tell if this *E coli* is causing diarrhea is to have your vet get the bacteria "serotyped" at a diagnostic laboratory. This procedure can identify the strain of *E coli*. Your vet probably wouldn't normally do this unless many of your calves are affected.

To fight this infection, vaccinate the pregnant cow 6 weeks and 2 weeks before calving. On subsequent years you need only give a booster dose 2 weeks before calving. The cow develops high levels of antibody to *E coli* and passes them in her colostrum. When the calf nurses, the antibody protects the calf's digestive tract against the bacteria. Again, you can see why it's so important that the calf get colostrum.

Another option for protecting young calves against *E coli* infection is to use a genetically engineered antibody. You give this monoclonal antibody to the calf immediately after birth. It will provide immunity to *E coli*. Either prevention method works quite well. Vaccinating the cow requires that you plan ahead, but it's relatively inexpensive. The monoclonal antibody is considerably more expensive.

Another disease that threatens young calves is infectious bovine rhinotracheitis (IBR). This viral disease can infect calves immediately after birth. It may cause diarrhea and pneumonia, and many affected calves die. The diagnosis of IBR is based on clinical signs and laboratory tests. There are several ways to protect calves against IBR. Most important, vaccinate your cow herd against IBR annually. This ensures that each cow's colostrum will contain antibody against IBR. Second, vaccinate the calf at birth with an intranasal product. Alternatively, you can give a killed-virus vaccine in the calf's muscle, but this doesn't offer solid immunity. Conversely, intramuscular modified-live vaccine gives reliable immunity. However, it may put too much stress on the calf, and vaccinated calves can shed the IBR virus to infect other animals in the herd. The virus incubates in a calf's nasal membranes and is spread through coughing.

A third type of infection in calves is bovine virus diarrhea (BVD). It rarely affects calves immediately after birth, but occasionally it occurs in calves born of heifers. You may think the calf was actually born with it. That can actually be the case with both BVD and IBR. The only way to differentiate one virus from another is for your vet to run lab tests. Both can cause diarrhea, pneumonia and death. Some researchers believe that a fetus can become infected with BVD virus in early pregnancy. In these instances, several complications can develop (see Chapter 14).

One way to fight BVD is to use a killed vaccine on the cow herd annually. While safe, this is not very protective. A second type of vaccine, the modified-live vaccine, confers excellent immunity but may cause virus shedding both in cows and calves. There's a lot of controversy over whether BVD virus shedding hurts unborn calves of herdmates. Until these controversies are resolved by further research, don't be surprised if your vet finds it difficult to give you a firm recommendation on how to vaccinate for BVD.

Another disease to which calves are usually exposed before weaning is pasteurellosis. It's caused by 2 strains of *Pasteurella*, either *Pasteurella hemolytica* or *Pasteurella multocida*. The first can be found in dairy calves but is much more common in beef calves. The second strain is more common in dairy calves. Both cause pneumonia in calves and in some instances can cause acute illness that leads quickly to death.

Generally, pasteurellosis is regarded as a pneumonia problem in conjunction with other diseases of the bovine respiratory disease complex. *Pasteurella* is a normal inhabitant of the upper respiratory tract of calves. We only consider it a pathogen when it infects the lungs. You could call this organism an opportunist because it multiplies when the calf is stressed by other viral infections or ventilation problems. Pneumonia caused by *Pasteurella* is usually more severe when it occurs with IBR or BVD infection. Nearly every calf is exposed to *Pasteurella* many times in its lifetime. Your vet will base a diagnosis of pasteurellosis on clinical signs and positive confirmation by culture of lung tissue.

The best way to prevent pasteurellosis is to upgrade ventilation of your calf housing and use vaccines. Killed vaccines are controversial. Many experts believe they actually make cattle more susceptible to *Pasteurella* pneumonia. Research has provided us with a live intradermal *Pasteurella* vaccine. Though experts still argue the merits of a live vaccine, work has shown it to be highly effective in lowering the incidence of pneumonia in beef calves. Neither vaccine has been widely used in dairy calves.

Another infection to guard against is rotavirus infection. This virus can cause acute septicemic diarrhea that usually kills the calf in 24 hours. Rotavirus infections occur primarily in the first 3-5 days of life. The calf dehydrates rapidly and goes into shock. A positive diagnosis depends on clinical observation and lab tests to isolate the virus.

The most effective way to prevent rotavirus infection is to give the modified-live oral vaccine. In most instances, this vaccine works well. There is more than one strain of rotavirus. Beef cattlemen have a vaccine to inject their pregnant cows so newborns will have protection when they nurse, but in dairy cattle the calf doesn't nurse long enough on the dam to maintain adequate protection.

Your calves may have been infected by coronavirus without your knowing it. These infections generally occur when the calf is 1-2 weeks old. This virus causes chronic low-grade diarrhea and sometimes pneumonia. The stool is black and may be slightly tarry. Calves gradually waste away, and some bloat. About half of infected calves die because they lose so much weight and are malnourished. In some herds about 10-20% of calves are infected. Your vet can diagnose coronavirus infection only by doing a postmortem examination and delivering a fresh piece of the intestine to a lab for virus tests.

An oral vaccine prevents coronavirus infection. It's usually available in combination with the rotavirus vaccine. Not everybody agrees on how effective coronavirus vaccines are.

Another infectious organism is *Hemophilus somnus*, which causes pneumonia and occasionally arthritis in calves 6 weeks of age and older. The pneumonia develops quickly and the calf soon has a fever of

104-105 F. Your vet can be sure it's *H somnus* by culturing some lung tissue from an infected calf. These cultures are difficult and your vet must handle samples in a special manner when sending them to a lab. I usually consider the possibility of *H somnus* infection when I'm called to investigate a pneumonia problem in a cow herd. A killed vaccine is available to prevent this infection. It usually takes 2 doses, given 2 weeks apart, to provide good protection.

Clostridium infections in calves occur mostly as blackleg and overeating disease (see Chapter 14).

Another optional vaccination is for Johne's disease (paratuberculosis). If Johne's disease is a problem in your herd, your vet may suggest vaccinating calves 30 days after birth.

From Weaning to Breeding

Feeding

You can prepare calves for weaning by offering them a "starter" ration as early as their second week of life. This feed should be highly palatable, high in protein (15-20%), and low in fiber. Let calves have all they want of it, up to 4 lb per day. When the calves are 4 months old, you can replace the starter with a heifer grain mix. Calves can digest hay, haylage and silage earlier than you may think. I suggest you offer hay and pasture as forage for calves younger than 4 months old because silage heats up and can become moldly in the manger. The best forage for a young calf is usually legume grass hay. Some calf starters even contain forages. In that case, you need offer only the starter and no additional forage. Be sure to keep the forage percentage in the starter under 25% in these early months.

Though you may have read about the hazards of overfeeding heifers, I think animals suffer more from underfeeding than overfeeding on most farms. Replacement heifers tend to be forgotten animals. Heifers ought to be gaining well by the time they reach breeding age. They require 65% TDN (total digestible nutrients) in their ration until they weigh 600 lb (Table 1). Hay and hay silage may run 62-63% TDN but generally average closer to 58% on a dry-matter basis. Pasture seldom exceeds 60% TDN.

Because you can't meet the 65% TDN requirement with forage alone, add some grain or high-energy silages for heifers less than 1 year old. If you leave out the grain or high-energy silages, you end up lowering the intake and digestibility of the forages. This leaves too little energy available for the rumen bacteria to aid digestion. Good corn silage that averages 68% TDN would be a good supplement. But it's too low in

Table 1. Daily nutrient requirements for growing large-breed heifers.

Body Weight (lb)	Age (wk)	Daily Gain (lb)	Crude Protein (lb)	TDN (lb)
110	3	1.10	0.63	3.20
220	14	1.76	1.06	4.64
330	23	1.76	1.40	6.00
440	32	1.76	1.75	7.34
550	41	1.76	1.88	8.64
660	50	1.76	1.94	9.90
770	64	1.76	2.16	11.30
880	78	1.76	2.50	12.69
990	88	1.76	2.85	14.20
1100	98	1.76	3.25	15.70
1210	110	1.76	3.71	17.46

(Adapted from *Nutrient Requirements of Cattle,* 6th ed. National Academy of Sciences, Washington, DC, 1988)

protein to be the sole feed for these animals. I often suggest a heifer ration that's 1/3 corn silage and 2/3 good hay or hay silage. Here's a feeding schedule based on high-quality forages:

Birth to 12 Months Old: Feed alfalfa grass hay or low- moisture silage containing at least 56% TDN and 17.5% crude protein free choice. Also offer 3-5 lb of grain mix daily with 80% TDN and 14% crude protein.

12-24 Months Old: Feed 1/2-2/3 of the ration dry matter in alfalfa grass hay or hay silage. Another 1/2-1/3 of the ration can be corn silage. Always offer salt, minerals and water. Because heifers' primary feed is forage, they need a higher phosphorus mineral supplement than do younger calves. Use Table 1 to match specific nutritional requirements to the amount of weight the heifers gain.

Dietary protein and energy requirements for pregnant heifers are similar to those for open heifers. You should be aware that feeding according to book guidelines may not be adequate. Use your critical eye and body condition scoring to know if nutrient levels are sufficient in heifers. The mineral requirements for heifers are about the same as for dry cows except that there is no upper limit on calcium levels (see Chapter 19).

Vaccinations

From weaning to breeding, use the same types of vaccinations you give the cow herd. Starting at about 6 months of age, vaccinate heifers twice (2 weeks apart) for IBR, BVD and 5-strain leptospirosis (lepto). Repeat this procedure annually. Optional vaccinations include *Hemophilus somnus*, *Clostridium* and *Brucella*. Give the first *H somnus* vaccination when the heifer is about 4 months old. Give another in 2 weeks, then repeat annually. To prevent clostridial infections, use a 7-way product at 3-4 months of age. Give 2 doses 2 weeks apart, then give annual boosters. Some dairymen, especially those who plan to sell breeding stock later on, vaccinate against brucellosis (Bang's disease). You may earn as much as $50 more per head at selling time if your records show you vaccinated the animals against brucellosis. However, you must follow the vaccination rules established in your state; vaccination age varies widely from state to state. Also, the brucellosis vaccine must be given by your veterinarian.

Deworming

Deworming is often neglected; it probably wouldn't be if dairymen understood the effect of worms and other parasites on the younger stock of a herd. By the time a heifer becomes a mature cow, she usually has some immunity to worms. But heifers are very susceptible to a number of external and internal parasites. Regardless of the type, parasites reduce performance. Use the same deworming program on pregnant heifers as for open heifers.

Lungworms are a common parasite, and are most serious where the winters are mild and the summers are warm and wet. Lungworms thrive when there's moisture and moderate temperatures of 65-70 F. They can withstand cold of 40 F but rarely survive temperatures as low as 0 F. Lungworms are thread-like in appearance and 2-3 inches long. They live in the bronchi and smaller air passages of the lungs. Adult females lay their eggs in the respiratory tract; the heifer coughs them up and swallows them. The eggs hatch in the digestive tract, mature into larvae and pass out of the animal in the manure. Lungworms reproduce in large numbers. It's been estimated that one calf can contaminate a pasture with as many as 33 million larvae.

Other heifers pasturing with an infected one become infected by eating the infective lungworm larvae in the grass. The larvae enter the intestine and migrate through the intestinal wall to reach the abdomen and lymph nodes. By 3-5 weeks after infection, these larvae have migrated to the bronchi and become mature worms that lay eggs.

Lungworms most often infect cattle on pasture. Usually the outbreaks are most serious in heifers 4-10 months of age. July and August are the months of highest infection rates. Fields and water supplies become heavily contaminated from the manure of infected animals. From 7 to 8 weeks after infection cattle may expel adult lungworms from the bronchi as they develop immunity to worms. When a heifer has many lungworms you may see frothy mucus around the mouth and nostrils. In milder cases, you may only see coughing and poor performance. Cattle with lungworms are susceptible to bacterial infections, and they may develop pneumonia and a fever of 104-105 F. Heavily infected heifers may suffocate.

Your vet can diagnose lungworms by microscopic inspection of a manure sample for larvae or by postmortem examination of a dead heifer.

Keen management is the best way to control these parasites. Put young cattle only on well-drained pastures. Also make sure they have high-quality feed and that their drinking water isn't contaminated. Good sanitation and housing are also helpful.

Because heifers can expel adult lungworms, some animals may recover without treatment. For animals requiring treatment, several drugs are available. Levamisole is highly effective. Treatment of the pneumonia may also be necessary.

A number of digestive-tract worms can infect cattle. Keep in mind that worms don't multiply within the animal. A calf with worms only acquires more worms by picking up more larvae on pasture. The worm that does the most damage to heifers is *Ostertagia*. It damages the lining of the abomasum and impairs digestion. *Osteragia* is common throughout the country, and it may be the most common parasite in cattle. It's representative of a large group of stomach and digestive worms; it is a good example of the kind of parasites that impair production and can cause serious economic loss.

Deworming with Rhubarb

Many of my clients arrange to deworm their heifers when moving them. This was the case with Dave. He wanted to get his heifers dewormed through the chute as they came off the trailer. It was a sunny January day. The ice on the roads was turning to slush, which made driving hazardous. I was to be at Dave's at 1:00 PM, but was hung up with a calving that needed tending at 12:30. My wife, Judy, had packed my lunch and I was gulping it down as I drove. Judy had packed some rhubarb stew. It was delicious, and I was having trouble eating it and keeping the truck on the road. I finished just as I got to Dave's.

As I got out of the truck, Dave asked what I was planning to use for a wormer. I thought the timing of his question was great, because I had been briefed on new deworming practices at a recent veterinary meeting. I began to talk about Ostertagia, *the medium-brown worm now thought to be of major economic importance in cattle.*

Two types of *Ostertagia* are recognized as more serious problems than once thought. Type I has a 21- to 28-day life cycle. The adult worm resides in the abomasum and feeds on the lining. It lays eggs that are passed in the manure. The eggs hatch in warm weather into larvae that molt several times. The larvae crawl from the cow patty onto a blade of grass. There they wait for another heifer to eat the grass, and thereby gain a free ride into the abomasum of a new host. The larvae burrow into the abomasum lining and damage the gastric glands. In about 4 days, they mature into adults, which lay a new batch of eggs. Most of the modern dewormers, such as levamisole, morantel, ivermectin and fenbendazole, do a good job removing the adult Type-I *Ostertagia.*

I told Dave the type-II Ostertagia *was much more serious. Dave had been intently staring at my coat, waiting to get a word in edgewise. When it looked like I was letting up for a breath, he interrupted. "Doc, what you been doin'?" At the same instant he asked, I saw what he had been staring at. In my haste to gulp my lunch, I had spilled a big hunk of rhubarb on the zipper of my coat. I knew what he was going to ask next, and in the same instant I knew that he hadn't been listening. I said, "Oh, nothing much but cleaning a couple of cows. Why?" Dave pointed at my coat, wrinkled his nose and said, "You got a gob on your coat." I quickly replied before he had time to think, "No problem" and I put my coat to my mouth and licked the rhubarb off.*

Dave's reaction was beyond my expectation. He turned 3 shades of green and went over and leaned against the barn. I thought he was going to lose his lunch. It seemed like an eternity before he could say anything. I couldn't keep a straight face. I had to walk around behind the truck. Dave finally figured out he had been "had." Both of us had a good laugh when I told him that I had been eating rhubarb.

Type-II *Ostertagia* larvae can lie dormant in the abomasum. These arrested larvae do no damage to the abomasum while dormant and hibernate for up to 18 weeks. They become dormant during certain weather conditions. When weather conditions again become ideal, they burst out of the gastric glands of the abomasum and quickly mature to egg-laying adults. Sometimes the adults produce eggs in waves and leave large numbers of larvae on the pasture.

Either way, the damage to the heifer's abomasum and health is devastating. In some instances, the heifer could die from the large number of larvae developing into adults. At best, infected heifers become "poor-doers."

In northern states, larvae may become dormant in the fall and winter. In late February and March, the weather begins to moderate, so the larvae mature to adults and start laying eggs. If a heifer is loaded with dormant larvae, their awakening in the spring causes serious damage to the abomasum. In the southern states, larvae may become dormant during the summer, when it is hot and dry. In the fall when the rains start, the larvae awaken from their dormant state.

By now Dave's color had returned to normal and he was interested. I told him that it was very important to select the correct dewormer. Conventional dewormers, such as levamisole or morantel, would only kill the egg-laying adults. Neither would kill the larvae buried in the abomasum.

It is important to use a dewormer such as ivermectin, fenbendazole or oxfendazole (which isn't yet approved for dairy cattle) that kills the dormant larvae before the larvae emerge. In northern climates, this would be in late winter. In the southern United States, these dewormers should be used in August and September.

Dave's face lit up as he said, "Golly, Doc, now I know why my heifers got so poor in the early spring last year. I was using levamisole, which only kills adult worms."

I told Dave that, while this was likely, he needed to do more with the deworming program. When turned out on pasture, heifers with heavy worm burdens would contaminate the pasture with worm eggs. In addition, many Ostertagia *larvae might survive the winter by burrowing into the soil a couple of inches.*

Not only should Dave deworm heifers to avoid contaminating the pasture, but he needed to be aware that young heifers could pick up a heavy load of worms in just a couple of weeks. It was very important that he deworm his heifers after about 3 weeks and 6 weeks on pasture to eliminate a new worm load. Dave asked me if I had stock in a dewormer company. He would be deworming his heifers all the time. I told Dave that I wanted to show him how to maximize his return on his heifers. Research has shown that deworming *heifers on worm-laden pastures would return Dave nearly $100 per heifer in increased performance.*

Dave also needed to work toward having clean pastures. He could avoid some of the deworming by turning his young stock on pasture that had been previously in crops and seeded to pasture. His older animals would have much more immunity to worms, so he could use the contaminated pastures for his older heifers. Until Dave could get organized with clean pasture, I suggested he deworm his heifers with ivermectin in the fall. This would kill the adult worms, dormant larvae and cattle grubs, and also help control lice. I also suggested that he deworm heifers in the spring, after they had been on pasture for 3 weeks and again at 6 weeks. Morantel recently became available as a sustained-release bolus so that deworming only need be done at turnout.

Morantel is now also available in feed form. This means that Dave can worm heifers at 3 and 6 weeks after turnout without having to catch his heifers. I advised Dave that morantel is ineffective against dormant Ostertagia *larvae, however. If he used another method for fall cattle grub control, an alternative to ivermectin would be to use fenbendazole for deworming in the fall also.*

If Dave's cattle were pastured in the southern states, he would need to deworm for dormant *Ostertagia* larvae just before the larvae emerge in August and September, rather than in late winter. Cattle on southern pastures are dewormed also at 3 and 6 weeks after turnout with sustained-release morantel or feed-grade morantel.

I told Dave to remember to use dewormers that killed dormant larvae or the rhubarb would be on his heifers in the spring.

Extra Teats

Extra teats are commonly seen on nearly all breeds of dairy cattle and are considered to be hereditary (inherited). Extra teats should be removed at the time of dehorning because the extra teat will have a small amount of milk-secreting tissue connected to it when these heifers freshen. This mammary tissue starts producing milk. It often develops mastitis and gives the udder an asymmetric appearance. While mastitis in the extra teat is usually not life threatening, it can be difficult to resolve.

Extra teats are relatively easy to remove if done before the heifer is "springing." Your vet can do this or he can teach you the technique. The heifer is restrained with a halter or in a head gate. Someone must hold the tail over the back to prevent the surgeon from being kicked. After the extra teat is scrubbed with an alcohol pad, it is stretched away from the udder so it can be snipped with scissors at skin level (Fig 1). It is wise to make the cut in the same plane as the folds of the udder skin so the scar will be less apparent. If the wound at the removal site gapes open, a suture should be placed. Follow the teat removal with a disinfectant or antibiotic spray. Occasionally I get a request to remove an

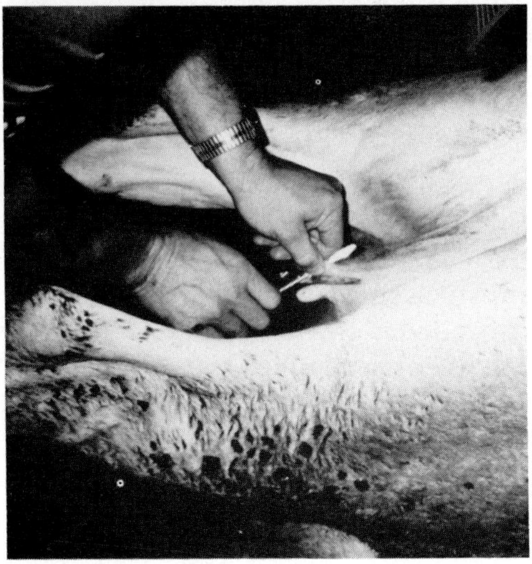

Figure 1. Extra teats on replacement heifers should be removed when the heifers receive their calfhood vaccinations. After disinfecting the area where the extra teat is to be incised, use sharp scissors to remove it. For a better appearance, make the teat excision parallel to the folds of the rear udder attachment skin.

"I've been out with Daddy, cutting off extra teats!"

extra teat on an adult cow. I only do these during the dry period so that I don't have to contend with milk seeping from the incision site. Teats on adult cows are usually much larger. An emasculator is applied to the base of the extra teat before removal. This minimizes hemorrhage and closes the wound so that sutures may not be needed.

Dehorning

Dehorning calves is less stressful if done at an early age. Mature cattle are always stressed by dehorning because their horns contain bony tissue. Cutting this bone is very painful unless the horn is deadened with a local anesthetic. In older cows, there is even a small risk of fracturing the skull when dehorning unless a saw or surgical wire is used.

Small calves should be dehorned with an electric dehorner (Fig 2). By 1 month of age, horn buttons can be felt. The calf is restrained and the hot iron is put around the horn button. Generally, a hot circular iron with a cutting edge is best. The iron is held to the horn button until the cutting edge gets down to bone. A rotating motion and a deft flick of the dehorner usually flip the horn button off.

When an electric dehorner is used, fly strike isn't usually a problem, so this procedure can be done year-round. Calves don't seem to be stressed nearly as much as by some of the other methods of dehorning. You must be careful with the hot iron because you can burn yourself or the calf other than its horn. This method also requires electricity.

Caustic paste is sometimes used on young calves just as their horn buttons are emerging. The caustic is vigorously rubbed on the horn button until a few drops of serum exude from the horn surface. This method works as well as the electric dehorner as long as the paste is rubbed in vigorously. If the application is faulty, the heifer will develop a scur or partial horn. This method is somewhat more hazardous if the caustic gets into the calf's eyes. This can happen when a treated calf stands out in the rain. If you are aware of this problem, it can be avoided. Caustic can also be hazardous to the operator.

For larger calves (200-600 lb), an electric dehorner or caustic powder is not satisfactory. These methods won't destroy all of the horn tissue. In these cases, a gouge is used. The size of gouge used depends upon the size of the horns to be removed. Dehorning with a gouge works quite well if a ring of skin and hair surrounding the horn is also removed with the horn. If the ring of skin and hair is not complete, a scur will grow back at the removal site. Bleeding is always a complication of dehorning with a gouge. The hemorrhage control depends upon the size of the horn and degree of bleeding.

While a nicely shaped head is often the result, gouging horns stresses a heifer. The stress is greater if the horns have grown to the point where they develop a horn sinus. On small heifers, the horn hasn't developed an opening into the frontal sinus. As the heifer gets closer to 6 months of age, the sinus connection forms. When an older heifer is dehorned, the inside of the skull (sinus) is opened to the environment. After dehorning, dirt and cold air can enter the sinus, often resulting in infection or sinusitis. These heifers don't eat well and can develop a pus-filled sinus that exudes pus onto the poll. Sometimes these infections lie dormant for months. Treatment usually includes flushing the sinus with peroxide and infusing antibiotics in the sinus with a dry-cow mastitis tube. Systemic antibiotics are also warranted.

Another method of dehorning is with the Keystone dehorner. The Keystone dehorner works like a guillotine on the horn. It is used on larger, more mature cattle. As with the gouge, the Keystone dehorner can also cause complications. A horn sinus is almost always involved. Hemorrhage is severe, and the pain and resultant stress are much greater. This method of dehorning should be avoided if possible. If a Keystone dehorner must be used, get your vet involved. On mature cattle with horns, sometimes sawing the horns off with surgical wire or a bone saw is indicated. This is best done under anesthesia by your vet.

After dehorning, any bleeding from spurting arteries can be stopped by pulling the arteries out. Oozing of blood from the dehorning wound should be stopped with styptic powder. If the horn sinus is open, use some clean cotton to plug it. This will protect the sinus from dirt and moisture. The cotton should not be jammed in so tight that it will not dislodge as the calf heals. Ideally, the cotton will be incorporated into the scab and fall off with the scab.

Figure 2. Using an electric dehorner is less traumatic to calves than gouging horns out. Today, many vets prefer to administer a sedative/anesthetic to a calf before dehorning. While firmly restraining the calf, burn the horn button until the skin around the horn turns a deep bronze color. When an electric dehorner is used in the summer, fly strike is seldom a problem.

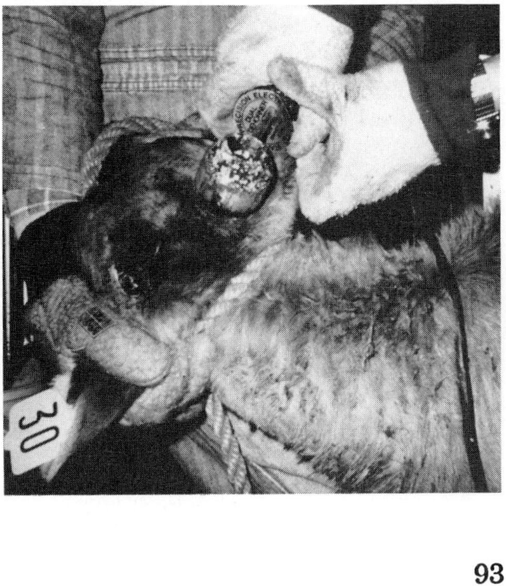

Rarely, bleeding cannot be stopped after larger bull calves are dehorned. In these cases, I often bandage the head. You may need to bandage calves that start bleeding after your vet has left. Put cotton over the wounds and use elastic or duct tape to press the cotton tightly on the wounds. Wrap the tape around the calf's head as though he had a tooth ache. After pressing the tape down firmly so it sticks to the hair, cut the band of tape under the chin so it does not completely encircle the head. If you don't, the head may swell and impair breathing, which could prove fatal.

Body Condition Scoring

Body condition scoring is simply an estimate of the quantity of fat under the skin in certain areas of a cow. The external fat is an indication of body fat reserves. Heifers ready to freshen should have a score of 3-3.5. If their score is higher than this, internal body fat may interfere with calving.

You may have a natural instinct for judging heifer body condition. Or you must develop the knack of knowing when your heifers have enough flesh or too much. Heifer body condition can be scored as follows:[6]

Condition Score #1: The wings of the vertebrae are visible over the area of the loin. Their lack of covering makes the heifer appear to have a shelf in the loin region. The hooks are sharp, with little flesh covering. The area below the tailhead and between the pin bones is depressed. The bones are sharp with little flesh covering. There is a severe depression between the hooks and pin bones.

Condition Score #2: The wings of the vertebrae are still visible in the loin area but are not sharply protruding. There is not a distinctive shelf appearance to the loin area. The hooks are prominent. The area below the tailhead and between the pin bones is depressed. The pin bones are prominent and the depression between the hook and pin bones is less pronounced.

Condition Score #3: The wings of the vertebrae in the loin area are only detected by feeling for them. As a group, the wings of the vertebrae in the loin area appear smooth, and there is no overhanging shelf appearance. A slight depression is visible in the loin. The hooks are rounded and smooth. The areas between the pin bones and around the tail head appear smooth without signs of fat deposit. The bones are rounded and smooth.

Condition Score #4: The wings of the vertebrae in the loin area can still be felt. They appear flat or rounded. The loin appears flat. The hooks are rounded and the span between the hooks is flat. The area around the tail head and pin bones is rounded, with evidence of fat. The pins can only be felt with firm pressure.

Condition Score #5: The wings of the vertebrae in the loin area and the hooks aren't visible, and fat is apparent. The tail head appears to be buried in fat. The pin bones aren't visible, nor can the pins be felt with firm pressure.

Breeding Heifers

Breeding heifers is usually easier than breeding cows if the heifers have had adequate feed. The first 2-3 heat cycles are usually much shorter than those of mature cows. If you watch closely, you may see heifers come in heat every 14-17 days. You should skip the first couple of heats until the heifer has a consistent cycle. Many heifers have a regular cycle of 17 days.

Heifers respond much more reliably to prostaglandin injections than do mature cows. They come into heat about 60-80 hours after prostaglandin injection if it is given during the luteal phase of the cycle (see Chapter 10). This is why heifers are often preferred as recipients in embryo-transfer programs.

Take special pride in your heifer breeding program. Your heifers represent the advances of modern genetics. If you have been using artificial insemination, their sires are representative of the latest bloodlines available. Your heifers represent years of advances in dairy science. They have genetic promise, so carefully select the bulls you use on them. First and foremost, select AI bulls that are easy calvers. This rating is usually available in the AI stud books. If you aren't using AI, at least use a bull that you *know* sires a small calf.

References

1. *J Dairy Sci* 63:973-977, 1980.
2. *JAVMA* 174:581-589, 1979.
3. Ibid.
4. *Vet Clin No Am (Food Animal)* 3:622, 1985.
5. *Hoard's Dairyman* No. 18, 1981.
6. Braun K: Personal communication, 1986.

9

Managing Fresh Cows

During pregnancy, a cow provides the nutrient requirements of her unborn calf. She supplies the calf with optimum levels of nutrients even at her own sacrifice. As a necessary part of her pregnancy, a cow has high levels of hormones in her system. She is a finely tuned machine that is highly susceptible to external stress. The combined effects of the demands of pregnancy and high hormone levels impair her natural defense mechanisms. For these reasons, you must provide newly fresh cows with special care to obtain optimum performance.

The First Week After Calving

During the first week after calving, see that your newly fresh cows get enough to eat and that they receive optimum dry matter intake. If a cow is not feeling well, give the necessary treatment so that she eats properly. Her feed must have adequate levels of fiber, so offer her plenty of dry feed. Limit high-moisture feeds, such as haylage, high-moisture corn and corn silage. Regulate dry matter intake so she receives an increase of 1 lb of dry matter each day. You can use vitamin B_{12} to sharpen her appetite if it seems sluggish. At the same time, see that she doesn't overeat on concentrate.

The first week after calving, when you're giving the fresh cow special attention, is an excellent time to switch her from the prefreshening ration to the milking ration because it's easy to monitor her dry-matter intake carefully.

A cow may only eat 20-25 lb of dry matter the first few days after freshening. This is normal, and is only half her daily consumption once she recovers from calving. Balance her ration so the concentrate to forage ratio is about 50:50. Some dairymen mistakenly balance forages to concentrates on the assumption that she's eating up to her full potential of 45-50 lb when in reality she's eating only 25 lb of dry matter. You may end up giving her far too much grain this way, and it throws the ratio off as far as 90:10 concentrates to forage. This is why many fresh cows have digestive disorders.

Newly fresh cows usually are fatigued; their hind legs may be wobbly and weak. Try to eliminate competition with the rest of the herd for food and water. If newly fresh cows must compete, they can easily be injured. Cows coming in heat and cows competing at the feed bunk often prey on newly fresh cows because they are vulnerable. A new "pecking order" must be established when these cows come back into the milking string. A newly fresh cow may be subjected to various types of stress before she gets her strength back. The result may be a lacerated, bruised, mastitic udder, or the "splits" and pulled tendons.

Milk fresh cows dry immediately after calving, especially cows with udder edema. Make sure they're milked completely and that they don't have udder problems. Research indicates that cows milked dry have a lower incidence of mastitis. These cows also give about 15% more milk for the entire lactation. The milk-producing cells (alveoli) are under pressure because of swelling at freshening, and they may stop producing milk. When internal udder pressure is reduced by milking, the resting alveoli begin producing milk again. The result is more milk for the entire lactation.

The first question you may ask is, "Will this practice give the cow milk fever?" The answer is, if she is going to get milk fever, she's going to get it whether she is milked or not. Even in Jerseys, incidence of milk fever does not increase when the cow is milked dry at the first milking.

Cows and heifers may have udder edema extending from the udder forward to the navel. Medication helps remove the excess fluid. Diuretics for this purpose include Naquasone, Lasix and Diuril. Naquasone is a combination diuretic and cortisol analog, which does a good job of reducing edema, but the cortisol portion may cause some problems. Cortisol suppresses immunity and in larger doses may cause a cow to go dry. It also depresses production of natural cortisol, which may interfere with the delicate balance of lactation hormones.

Lasix is a powerful diuretic that removes edema effectively, but it is more expensive than Naquasone. Administer it by injection or bolus. Diuril is not as effective as Naquasone and Lasix, and few dairymen use it nowadays.

Some dairymen like to use cortisol analogs, such as Azium, Meticorten and Predef. These drugs remove udder edema and stimulate appetite. But they also suppress immunity, which makes the cow more susceptible to infections, and they may cause cows to dry up. Follow your vet's advice on dosage and administration of these drugs.

Fresh cows are especially susceptible to environmental heat. In the event of infection, they may have higher fevers than usual. Even in the absence of disease they may develop fever. Improve their comfort by providing shade and free access to water. Try to provide a fly-free place where they can lie down and stay cool. Extra dietary potassium can be helpful, too. You may have been cautioned that too much potassium can cause udder edema, but there are instances where added potassium is beneficial. Consult your nutritionist for specific recommendations.

The prolonged spells of hot weather prevalent in such areas as the Southwest cause stress. Misting systems may help fresh cows withstand heat. This equipment sprays fine droplets of water over the cows for evaporative cooling (Fig 1). In areas of high humidity, however, this procedure may predispose cows to respiratory problems.

Keep fresh cows separate from the herd. Grass lots are excellent housing for fresh cows because they provide good footing and no competition from the milking herd. Where grassy lots are not available, dirt lots are acceptable as long as fresh cows don't have to compete for feed and water. The best way to maintain a lot for this purpose is to scrape away the upper layer of soil and replace it with clay as manure accumulates.

Problems in the First 40 Days

Milk Fever

Various illnesses may affect cows in the first 40 days after calving. One of the more common diseases is milk fever. The signs of this condition are quite variable. However, all affected cows have a low blood calcium level that causes gradual paralysis of various muscle groups. Usually the first muscle groups affected are the pupillary muscles of the eyes and the smooth muscles of the internal organs. If you examine the

Figure 1. These cows are standing under a misting system that cycles hourly when the temperature is above 85 F. The system keeps the cows much more comfortable at the feed bunk in hot weather. After installation, dry matter consumption increased and production improved accordingly.

eyes with a bright flashlight, the pupils either remain dilated or close very slowly. The smooth muscles of the digestive tract are also paralyzed, which causes constipation. If the cow has not calved yet, her uterine muscles will also be paralyzed. You may witness a slow delivery that results in a dead calf.

In the early stages of milk fever, a cow tends to be hyperexcitable. She may resist being caught, run into gates, or occasionally charge the person trying to handle her. The rectal temperature is depressed, rather than elevated. It may be as low as 99.5 F. Later, as the milk fever progresses, she becomes depressed and uncoordinated. The skeletal muscles eventually fail to function and she can't get up, even though she continues to struggle. She may get trapped under a gate or free stall.

Gradually she becomes more depressed and unresponsive to her surroundings. She develops an "S" curve in her neck and lies with her head to her side. Eventually her breathing becomes more labored and she may not be able to belch. As a result, she bloats and may die. In other cases, the cow stretches out on her side. Gas and feed from the rumen regurgitate into the throat, nostrils and windpipe. When she inhales, she may aspirate the feed into her lungs. As a result she may drown or, after recovery, may develop aspiration pneumonia. If the milk fever is not treated, she will die.

You can diagnose and evaluate the progress of milk fever by certain signs. Look for sluggish pupillary reflexes, excitability or depression, constipation, paralysis, bloat and abnormal body position. A red flag should go up if you observe many of these signs, and diarrhea instead

Figure 2. A simple way to diagnose milk fever or toxemia in a cow is to check the response of the cow's pupils to a bright light. The pupil at left is that of a cow with toxemia. It is tightly constricted before shining a bright light into it. The pupil at right is that of a cow coming down with milk fever. It either will not constrict in response to the light or will constrict slowly and only slightly.

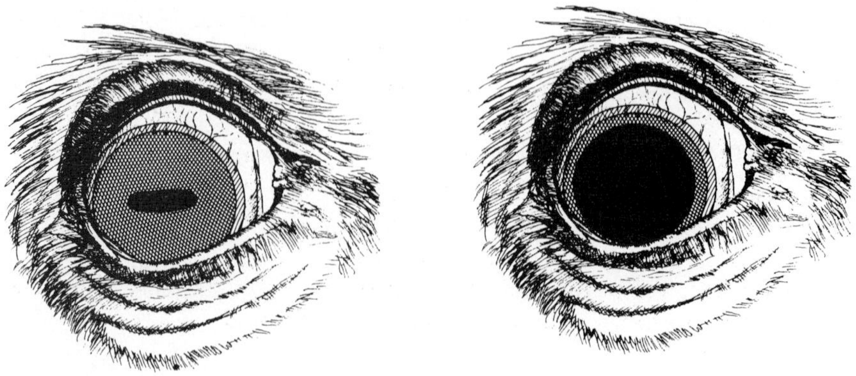

of constipation. In these cases, the cow may have another condition, such as mastitis or toxemia (Fig 2).

Milk fever is treated with intravenous calcium. Various calcium solutions are available. I give calcium at 1 g/100 lb of body weight. Most large-breed cows require 2 bottles of calcium intravenously. You may be qualified to treat your cows for milk fever, but be sure to follow your veterinarian's advice. When you give calcium this way, you raise the cow's blood calcium levels to extremely high levels. Normally, the calcium is distributed to the various muscle groups and paralysis quickly disappears. But if the calcium is given too quickly, the cow may develop heart irregularities and die suddenly. Other drugs can be used to reduce the shock of calcium to the heart, but occasionally these cows die in spite of the vet's best efforts.

Every cow maintains a balance of calcium in her body; some comes from her feed and some is reserved in her bones. This balance is regulated by the parathyroid gland. Some dairymen think the intravenous calcium corrects a calcium deficiency. What really happens is that the intravenous calcium causes a rebound effect. It encourages the cow's parathyroid gland to regulate calcium. If calcium blood levels begin to drop after calving, the parathyroid gland usually causes calcium to move from the bones to the blood stream. If calcium levels are high in the dry-cow diet, no calcium is needed from the bones. The parathyroid gland rests because calcium is not needed from the bones. At freshening, the cow suddenly needs a large amount of blood calcium as a result of the higher calcium requirement of the calf and demands for milk production. She can't meet that requirement from her diet, so the parathyroid gland usually pulls calcium from the bones to meet this need. The parathyroid gland may fail to respond if it has been at rest. The result is low blood calcium and milk fever.

A cow's parathyroid gland is a bit like a diesel tractor in the winter. On a cold, wintry day, if you try to start a diesel and run it full throttle without warming it up, it may fail. On the other hand, if the diesel is warmed up, it functions effectively. A cow's parathyroid gland is like that diesel. The gland can be "warmed up" by keeping dietary calcium low during the dry period. This causes the gland to move calcium into the blood from the bones. When the demand for calcium increases at freshening, the parathyroid gland is ready to respond (see Chapter 19).

If your herd has a history of milk fever that fails to respond to intravenous calcium, you need to do some detective work. Ask your vet to take samples of blood from a few cows to check calcium and other essential mineral levels. You may discover that your cows have other metabolic disorders quite separate from milk fever.

Most cows respond quickly to intravenous calcium. Those that don't respond may develop complications, such as musculoskeletal injuries from lying on one side for too long. When a cow doesn't get up on her own within about 30 minutes of treatment, you should roll her over on her other side to help alleviate cramps in her down leg. Then roll her from side to side every 2-3 hours to reduce the likelihood of "compartmentalization syndrome." When a cow lies on one leg too long, the circulation is impaired. As a result, she gets leg cramps and eventual muscle damage (Fig 3). As you know, the cow depends heavily on her hind legs to get up. Big, heavy cows generally are more susceptible to muscle damage than smaller cows. You may consider retreating a cow if she doesn't get up in 4-6 hours.

Another common complication of milk fever is retained placenta. The uterus can become paralyzed as a result of low blood calcium during milk fever, and a cow is unlikely to expel the placenta if she hasn't passed it before getting milk fever. Cows that have already passed the placenta may still have paralysis of the uterine muscles, and the uterus can't contract down to its prepregnant size (a process called involution).

Immediately after calving, the uterus is quite relaxed. In addition, the pelvic canal is wide open, so the uterus may easily turn inside out and slide out through the vulva. This complication is called a prolapsed uterus. A cow down with milk fever and trapped under a gate may push the uterus out as she struggles to get up. Frequently, she'll go into shock; she'll die if you don't offer prompt medical care.

Cows that have had milk fever often develop other conditions. As a result of retained placenta and poor uterine health, they are more

Figure 3. This cow injured her left rear Achilles tendon while she had milk fever.

susceptible to cystic ovaries and fertility malfunctions. All of this stress on the cow can also result in a displaced abomasum.

Retained Placenta

Newly fresh cows often have other complications. Retained placenta is very common even in the absence of milk fever. You probably have seen a placenta dangling from a cow's vulva. It may hang all the way to the ground. Such cows usually have a foul odor and they subsequently develop a uterine infection.

The causes of retained placenta are many, but one of the most common is difficult calving (dystocia). A cow is likely to have a retained placenta if she has been assisted at delivery and somebody has inserted their hand into the vagina or uterus. Many dairymen assist in delivery and then forget that the cow is not likely to pass her placenta afterwards. A few days later the cow is gravely ill.

Another cause of retained placenta is poor nutrition. A deficiency in vitamin E and selenium makes the cow highly susceptible to the problem and, in general, to poor uterine health. Too much potassium fed before calving, as well as a wide dietary cationic ratio, may further dispose a cow to this problem (see Chapter 19).

As discussed more completely in Chapter 19, cows misfed during the dry period, fat dry cows, and cows with nutritional imbalances often retain their placenta. Retained placentas also occur when cows have systemic diseases, such as IBR, BVD, leptospirosis and brucellosis. Older cows are at more risk than younger cows in that uterine involution is usually slower in older cows because of scarring of the uterine lining.

A poor calving environment can also increase the likelihood of retained placenta. In the process of calving, a cow may draw bacteria into the uterus from the environment. This is especially true when sawdust or wood shavings are used for bedding. Retained placenta is also common after abortion or early calving. The toughest retained placenta cases are in cows that have aborted. The placenta adheres tightly to the endometrium in these cows. The endometrium hasn't had time to prepare for release of the placenta, so it is seldom readily expelled.

You can diagnose a retained placenta by sight, smell and sometimes by vaginal examination. If a fresh cow is off feed or running a fever, in all probability she has a uterine problem. Treatment depends upon whether the cow has a fever. A fever indicates systemic disease. These cows often are off feed, milking poorly and lethargic. They require systemic antibiotic therapy. Consult your veterinarian in these cases. Large doses of penicillin G are by far the most effective and economic treatment. Drugs designed to break fevers, such as aspirin or flunixin, may also be given. Flunixin requires a vet's prescription. Some cows with systemic involvement also require intravenous fluid therapy. In nearly every instance, it is wise to not remove the placenta manually. As a matter of fact, manually examining or instilling medication into the uterus disrupts the uterine immune system for up to 8 days in routine cases.

Cows without systemic involvement usually eat and milk well. In these cases, giving the cow a few days to pass the placenta may suffice. But don't forget to watch her closely for fever or being off feed. Many veterinarians use prostaglandins to hasten passage of the placenta. We know that prostaglandins interrupt the luteal phase and thereby remove the progesterone influence on the uterus. Progesterone makes the uterus very susceptible to infection. When the uterus is under the influence of estrogen, it is more resistant to infection. Years ago, vets used stilbestrol or ECP to stimulate uterine involution. Now we know that while the estrogen effect was desirable, the oil base caused an undesirable long-term estrogen effect. The result was cystic ovaries and Fallopian tube damage. Current information suggests that estrogen given in a water base gives a very desirable boost to uterine resistance without deleterious long-term effects.

Prostaglandins also work directly on the uterine caruncles or buttons to which the placenta is attached. Prostaglandins stimulate white blood cells to loosen the placenta. A Yugoslavian study recently demonstrated that cows with uterine infections given prostaglandin at day 19 after calving clearly outperformed herdmates that were given uterine infusions. The prostaglandin-treated cows had a 60% advantage in average days open, number of services to conception and interval to first heat. This therapy needs more research to provide answers on optimum dosage frequency and efficacy.

If a cow does not pass her placenta within a few hours after calving, another option is to give oxytocin. But researchers claim that oxytocin doesn't have an effect on uterine motility after calving is over. Though oxytocin does not improve uterine motility, it does enhance the uterus's ability to fight infection and the placenta-loosening process.

Some vets use a product called ergonovine to help a cow pass her placenta. This may only help in cows lacking uterine contractions, as may be the case with milk fever.

My recommendations for cows with retained placenta include using oxytocin soon after calving and then following up with prostaglandin and estrogen therapy. If the cow has systemic involvement, she should also receive systemic antibiotics and intrauterine infusions of dilute iodine. In any case, rely on your vet's suggestions on what is best for your herd.

Pyometra

Another problem occurring during the first 40 days after calving is pyometra. A cow with pyometra has a uterus distended with pus, and ovaries that are not cycling. Pyometra doesn't cause many signs. Common signs are missed heats and white pus discharging from the vulva after uterine involution should be complete. The causes of pyometra are many and varied. The most common is an infection in the uterus during the middle portion of the cycle. When a cow is in the high-progesterone part of her cycle, the luteal phase, her uterus is much more susceptible to infection. This is between day 6 and day 16 of the cycle. In contrast, when she is under the influence of estrogen in her cycle, the follicular phase, the uterus is much more resistant to infection.

A cow often has her first ovulation about 10-14 days after calving. After ovulation, she is in the luteal phase. If her uterus has not fully cleaned or has an impaired defense mechanism, conditions are right for development of pyometra. The irritated and inflamed uterus is susceptible to organisms that reproduce in great numbers. The uterus responds by secreting mucus and sending in white blood cells for defense. It interprets the accumulated fluid as a pregnancy; it isn't able to drain the infection because there's no chemical signal coming from the uterus to cancel the luteal phase. As long as the uterus is full of mucus and pus, this chemical signal will not be sent to the ovary and the condition will persist.

A number of conditions predispose cows to pyometra. Cows with poor uterine immunity are more prone, as are herds with the venereal disease trichomoniasis. Pyometra may be more likely in cows that are zealously treated with uterine infusions, especially during the luteal phase of the cycle. Try to limit metritis infusions. If you must infuse a cow's uterus, do so during the follicular phase of the cycle. Uterine infusions impair a cow's natural defenses, especially during the luteal phase.

Pyometra is diagnosed by rectal palpation. Because pyometra can easily be mistaken for pregnancy, the palpator should know the history of the cow so there's no confusion. If you're observant, you may detect the pus discharge from the vagina and note that the cow has not been in heat.

Treatment of pyometra must include prostaglandins. Occasionally I also give uterine penicillin infusions when the cow comes into heat. But infusions should be used only in connection with prostaglandin therapy. Used alone, they may just make the problem worse.

Septic Metritis

Another complication seen in newly fresh cows is septic metritis, a serious uterine infection. Affected cows often have a foul, bloody vulvar discharge and may be straining. They may also be seriously ill with endotoxic shock, caused by the poisons in the uterus. If the cow is not in shock, she may have a fever, but many affected cows in shock have no fever. Frequently, you'll also notice she's off feed, has diarrhea, and has begun to dehydrate. Milk production also drops off significantly.

Causes of septic metritis include retained placenta, trauma from a difficult calving, or calving in a dirty environment. The invading bacteria are usually *E coli* or *Clostridium*. A cow with an impaired defense mechanism is more susceptible to septic metritis, too. Herd infections or nutritional deficiencies cause such impairment. Keep your eye on fat cows and cows that deliver twins; they seem more prone to septic metritis.

Septic metritis may be obvious in very sick cows, but usually the condition is diagnosed by a veterinarian doing a systematic examination. This includes evaluating rectal temperature, pulse and respiratory rates, and the reproductive tract. Treatment of septic metritis must be intensive and includes antiprostaglandins, antibiotics and fluids. Sometimes the uterus must be flushed. Most of all, these cows need a lot of tender loving care to recover. Don't allow affected cows to run with the rest of the herd until they recover.

Fat Cow Syndrome

Cows with the fat cow syndrome have ketosis or go down with what appears to be milk fever. They often are called "creepers" because they crawl around and can't get up. At calving these cows appear overconditioned and have a body condition score of 5 (see Chapter 6). They have a poor appetite and usually have problems calving.

The cause of fat cow syndrome is too much energy intake during late lactation. This often occurs when a cow has a long history of infertility. Many have had a long dry period.

Cows fed corn silage free-choice during late lactation and the dry period also are prone to fat cow syndrome. Low protein levels occasionally are incriminated. If protein levels are lower than required, cows readily put fat on their backs instead of milk in the pipeline. Some experts think that low selenium levels in the diet make cows prone to poor liver function. It is the liver that must metabolize the extra fat. If it malfunctions, cows may develop fat cow syndrome. For detailed nutritional information related to this syndrome, see Chapter 19.

This syndrome is diagnosed from clinical signs of an overconditioned cow. If in doubt, your vet can do liver enzyme tests to confirm the diagnosis. I have sometimes verified the diagnosis on postmortem examination (autopsy). The liver of affected cows has a nutmeg appearance because of the fat in it.

Treatment has variable success. Drugs to protect the liver and stimulate its function are necessary. These include B vitamins, amino acids, vitamin E, selenium, insulin and glucose. The best results are seen in cows receiving excellent care. If fat cow syndrome is suspected before a cow becomes sick, feeding niacin is very helpful.

Ketosis

Ketosis, also called acetonemia, is often considered by experts to be a sign of a metabolic disease, such as fat cow syndrome or displaced abomasum, rather than a separate disease. A cow with ketosis often has a sweet smell to her breath, and her urine will be positive to a cowside test called Ketostix (Fig 4). The cow is usually off feed or eating poorly. The nervous form of ketosis is considered an emergency. These cows appear "hyper" and may eat dirt ravenously or chew on fences and parlor stalls. They can also go berserk and charge buildings or people.

Ketosis develops when a cow takes in less dietary energy than the energy she expends in milk production. The cow is burning fat off her back and is not able to eat enough to keep her energy levels in balance. A cow can readily get ketosis if she has been on an inadequate level of energy during her dry period. Also, any factor that causes a poor appetite may lead to ketosis if the cow is milking heavily. This includes displaced abomasum and retained placenta. Nutritional factors that can cause

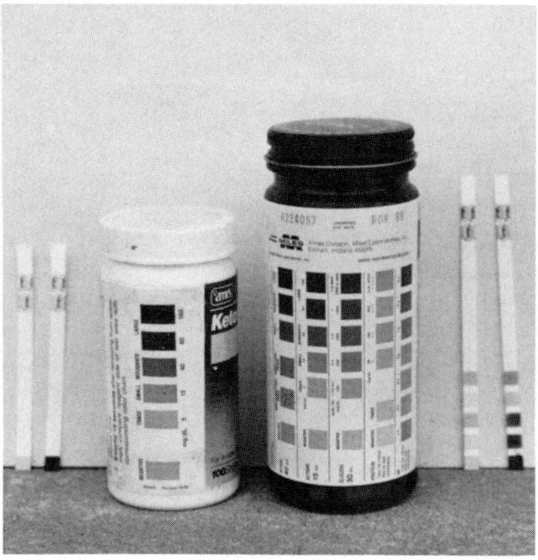

Figure 4. Cows can be checked for ketosis by dipping a test dipstick in the cow's urine. If positive for ketosis, the stick changes color from pale beige to deep purple. The stick on the left only checks for ketones, while the one on the right measures protein, pH and glucose, as well as ketones.

ketosis include low dietary magnesium, phosphorus or cobalt. The diagnosis is based on seeing typical signs of ketosis. Confirmation is made with the Ketostix test.

There are many treatments for ketosis. The fundamental part of most treatments is to give more energy in the form of intravenous glucose. Insulin, cortisone, niacin, amino acids and vitamin solutions also are commonly used. Sometimes intravenous fluids and bicarbonate are necessary.

The incidence of ketosis can be reduced by maintaining proper body condition with correct feeding programs and by body scoring. Nutritional programs must be correct to prevent deficiencies that may create the conditions for ketosis in your fresh cows. Overconditioned cows can be helped by feeding niacin late in the dry period and during early lactation (see Chapter 19).

Displaced Abomasum

Displaced abomasum occurs when the cow's fourth stomach compartment, the abomasum, slips out of position and becomes trapped, usually on the left side (Fig 5). The cow goes off feed and may become ketotic. She gradually becomes dehydrated and the manure usually is very dry, almost like mortar.

Many factors contribute to displaced abomasum. A low-fiber diet, especially during the dry period, may contribute to the condition. This is often compounded by stress from such conditions as retained placenta or milk fever. If the ration has a wide cationic ratio (see Chapter 19), a displacement may be more likely. Fresh cows that are getting insuffi-

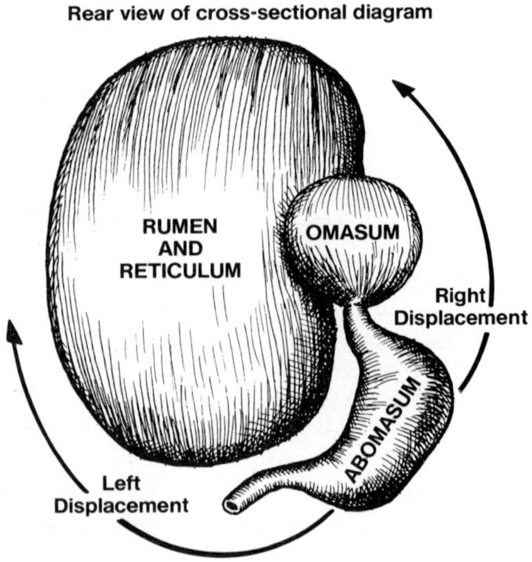

Figure 5. The abomasum, the fourth compartment of a cow's stomach, is vulnerable to sliding out of position to the left or right side.

Figure 6. The line drawn indicates the area where you may hear the tinkling sound of a displaced abomasum.

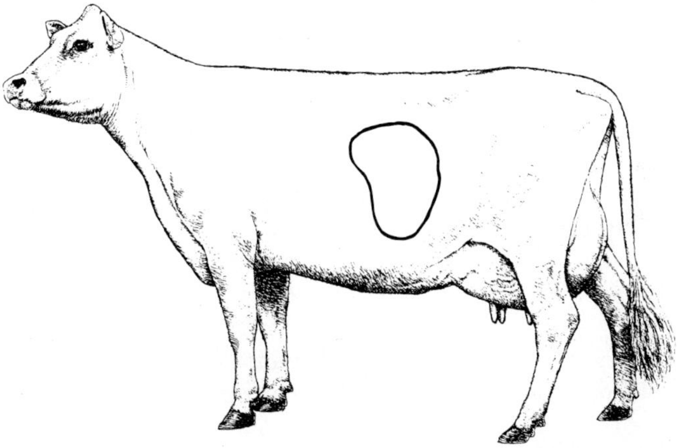

cient care are also more prone, such as cows fed high levels of grain without adequate forage. Low dietary selenium levels are also contributory. Low-selenium diets have an indirect effect by causing cows not to clean. Some researchers speculate that low dietary selenium may actually cause an unborn calf to develop a weak abomasum. These unborn calves later become susceptible as adults. If this is proven to be true, then selenium levels must be corrected in the pregnant dry cow if predisposition to displaced abomasum is to be prevented in the unborn calf.

The diagnosis of left displaced abomasum can only be positively confirmed by listening to a cow's rumen on the left side with a stethoscope and then "plunking" her side with your finger (Fig 6). The characteristic "ping" sounds like a banjo string or the plunking of a tin can. Sometimes the sound is loud enough to confirm without a stethoscope. This diagnostic technique is called auscultation and percussion.

Displaced abomasum is treated by getting the abomasum back into the normal position in the lower right portion of the abdomen. This can be achieved in several ways. Commonly, a surgical approach is used. Depending upon where your vet was trained, he or she may make the incision on the right side or the left side, or turn the cow on her back and use a ventral approach.

Cows often quickly respond to surgical correction and are soon back to nearly full production. Surgery has the added advantage of giving your vet the opportunity to explore and diagnose other complications in the abdominal cavity. A disadvantage of any abdominal surgery is the risk of infection. The surgery is often done in a dairy barn. If the facilities aren't clean and free of flies, or your vet is sloppy with his technique, postsurgical infections are common. Sometimes postsurgical infections are a problem even when the facilities are clean and the surgical technique is good.

A low-cost method for correcting abomasal displacement is the "blind stitch" technique. This involves putting the cow down with a reef rope and rolling her on her back. Your vet will then listen to see if the abomasum has shifted. When it is back in a normal position, a large needle with heavy suture is inserted through the body wall and into the abomasum, and then brought back out for tying. A variation of the blind stitch is to use a trocar with a plastic anchor to attach the abomasum to the body wall. The cost of this method is usually about a third to half of that of the more complicated surgical methods. As its name implies, the blind stitch method has the disadvantage of blindly stitching the organ in place. There is a risk of lacerating a milk vein and causing serious bleeding. Instances of puncturing the liver or intestine have also been reported. Despite these concerns, vets that routinely use this method seldom have complications. In the hands of the inexperienced, however, the risks are significant.

Some dairymen advocate rolling a cow to slide the abomasum back into position. This is satisfactory if a cow is going to slaughter shortly. But if you're just trying to avoid a surgery bill for a cow you plan to keep, rolling isn't satisfactory. The cow usually gets right back on feed, and then redisplaces a few days later. She can be rolled regularly until she gets straightened out. But by the time displacement is no longer a problem, the cow has been off feed enough to ruin her production level. She ends up being a mediocre producer for the rest of lactation.

The only other alternative is culling for slaughter. If treatment isn't done, the cow will either become a poor producer or die of starvation.

Chronic Endometritis

A uterine infection that has localized but has not become systemic may be termed chronic endometritis. These cows may have a discharge of pus from the vulva.

A cow with chronic endometritis is prone to cystic ovaries. Conversely, cystic ovaries make a cow prone to chronic endometritis. The aftermath of chronic endometritis often is infertility or resorption of the fetus.

Chronic endometritis usually results from complications of calving, such as retained placenta, acute metritis or pyometra. Some cows calve normally but have nutritional deficiencies that impair their natural immunity and the inability to eliminate an infection. Older cows are more likely to have impaired uterine defenses and subsequent endometritis.

Chronic endometritis is best diagnosed by rectal palpation by a competent palpator. A vulvar discharge of pus may make you suspect the condition, but diagnosis is best confirmed by palpation. A uterine culture can also aid diagnosis. Your vet must be careful interpreting the results because some cows have a positive culture and yet can still become pregnant. Culture samples should be obtained from a number of cows in the herd so your vet can look for a trend, rather than relying on results of an individual cow.

The usual treatment of endometritis is uterine infusion. Many chronic cases are caused by *Corynebacterium pyogenes,* which is best treated with 1-2 million units of penicillin infused into the uterus during heat. A second effective treatment is a sham embryo transfer flush. The cow is given the irrigation procedure just as in embryo transfer, but no attempt is made to program her or recover embryos. This procedure has worked well for me, but it is time-consuming and somewhat expensive.

Endometritis is best prevented by good nutrition, proper dry cow care and a clean calving environment.

The Case Against Uterine Infusions

Here are some other factors that should influence your decision to infuse cows. A reddish-brown bloody vulvar discharge during the first 10-14 days after calving is *not* a sign of endometritis but is a normal discharge of uterine fluid (lochia). This is highly desirable. A cow discharging pus may not actually have endometritis but may simply be clearing her uterus of infection.

Many of my clients conclude that pus from the vulva must mean an infected uterus. They never stop to consider that the pus may be coming from the vagina or the urinary bladder. Cows with healthy uterine defenses pass considerable amounts of pus after calving. Don't fall into the trap of thinking that vulvar discharge of pus means the uterus must be infused. It may only mean that the cow has good immunity and is cleaning herself up.

Considerable research data suggest that a cow is much better off not being infused. A cow's uterus is an airless environment that renders many antibiotics ineffective. Antibiotics that may not be effective in the uterus include gentamicin, neomycin and kanamycin. Also, many organisms in the uterus produce enzymes that neutralize infused antibiotics. For example, the bacterial enzyme penicillinase can neutralize most types of penicillin.

Pus and organic debris in the uterus inhibit many drugs, such as sulfas, gentamicin and kanamycin. Also, antibiotics are poorly absorbed into the deeper layers of the uterus immediately after calving. Therapy must be directed to the deeper structures, through systemic therapy, if there is any hope of eliminating the infection.

Antibiotic or disinfectant infusions of any type interfere with the action of white blood cells against infection. Even manipulating and handling the uterus upsets white blood cell function. Consequently, if you use an antibiotic that isn't effective and interferes with white blood cells, the results can be detrimental. Irritating drugs, such as furazolidone, oxytetracycline and Lugol's iodine, are notorious for impeding white blood cell function.

Equally important is concern about selling meat or milk from cows treated with antibiotics. No official guidelines exist for the effects of uterine infusions on milk withholding and slaughter residues.

Don't become unduly alarmed that there are no antibiotics left to use; there is a rational method of treating cows with uterine infections. It is based on the principle that a cow has considerable uterine resistance to infection during the estrogenic phase of the cycle. When the corpus luteum (CL) is eliminated, an estrogenic phase usually begins. The cow's uterus is very susceptible to infection during the progesterone phase of the cycle. Reduction of progesterone levels eliminates that inhibition of the uterine defenses. Increasing estrogen levels stimulate the white blood cells to attack organisms invading the uterus. Prostaglandins, such as Estrumate or Lutalyse, take advantage of a cow's ability to clean her uterus during the estrogenic phase. Prostaglandins dissolve any CL that is producing progesterone, and also stimulate the uterine mechanisms to expel any debris.

In a study of over 200 cows with uterine problems, half of the cows received a uterine infusion at 19 days after calving, while the other half received an injection of PGF_{2a}. There was no detectable difference in the rate of uterine recovery or involution. But PGF_{2a}-treated cows had a 22% better first service conception rate (34.6% vs 16.2%) and had fewer services to conception (2.42 vs 2.90). As you can deduce, the calving interval was also improved with PGF_{2a} treatment (89.5 vs 105.9 days). It is important to breed cows by 100 days if a 12-month calving interval is to be maintained. Sixty percent of the PGF_{2a}-treated cows achieved this goal versus only 38% of the infused cows.

Routine infusions after breeding may actually lower conception rates. If infusions do have a beneficial effect, it is likely that they first cause uterine damage, which then stimulates prostaglandin release and subsequent beneficial healing.

10

Detecting Heat and Breeding Cows

Dairymen tend to spend less time watching for heats and breeding cows than on any other management chore on their farms. Yet nothing is of greater importance in milk production than ensuring that every cow gets pregnant each year.

You need to know the basics of the estrous (heat) cycle to detect heat and breed cows at the right time. The cow has a 21-day cycle, with a 12- to 14-hour period of receptivity for mating. For purposes of definition, call the day of heat day 0 and day 21 the day before her next heat. On day 0, the cow is receptive to mating, and by day 4 her uterus is prepared for pregnancy. If conception does not occur or the fertilized egg dies, on day 17 the uterus and ovaries prepare for another heat.

Hormones of the Heat Cycle

Various hormones interact as the heat cycle progresses from day 0 to day 21. Each sex hormone will be discussed individually, but the different hormonal events occur simultaneously (Fig 1).

Estrogen

Estrogen is the hormone associated with heat. It causes the cow to become receptive to the bull. Signs of heat include vulvar discharges of mucus, restlessness, bellowing and mounting other cows. Estrogen also initiates development of the duct portion of the mammary gland in young heifers. This hormone is secreted by the growing follicles in the ovary. Estrogen levels are generally low when progesterone levels are high, and *vice versa*. At day 0, estrogen levels have peaked and the cow is showing heat. At the end of heat, estrogen levels drop rapidly and reach a low level by the time of ovulation. Estrogen levels remain low until day 17. Then the next follicular growth phase starts in preparation for the next heat.

Progesterone

Progesterone is commonly referred to as the hormone of pregnancy. It is secreted by a temporary structure on the ovary called the corpus luteum (CL). The CL is formed after ovulation and occupies the space of the ovulated follicle (Fig 2). The cavity left in the ovary at ovulation fills with a blood clot that organizes into a CL by day 4. At day 4, the CL begins to produce progesterone. If the cow conceives, progesterone levels stay high to sustain the pregnancy. Progesterone also quiets uterine muscle activity, stimulates uterine gland secretions, and makes the uterus an optimum environment for the embryo. The cow usually does not come into heat during pregnancy.

If the cow does not conceive, progesterone levels remain high only until day 17, when another follicular growth phase starts on the ovary (Fig 3). Progesterone is the predominant hormone between day 6 and day 16 because of the CL. This period of a cow's cycle is called the luteal phase. Progesterone stimulates growth of the glandular portion of the mammary gland in young growing heifers.

Luteinizing Hormone

Luteinizing hormone (LH) causes ovulation (release of the egg from the ovary). LH is released for a short time, then plummets to low levels.

Figure 1. Releasing hormones or releasing factors from the hypothalamus, located just above the pituitary gland, triggers the pituitary to produce hormones that regulate many systems, such as the thyroid gland, adrenal gland, kidneys, mammary glands and ovaries.

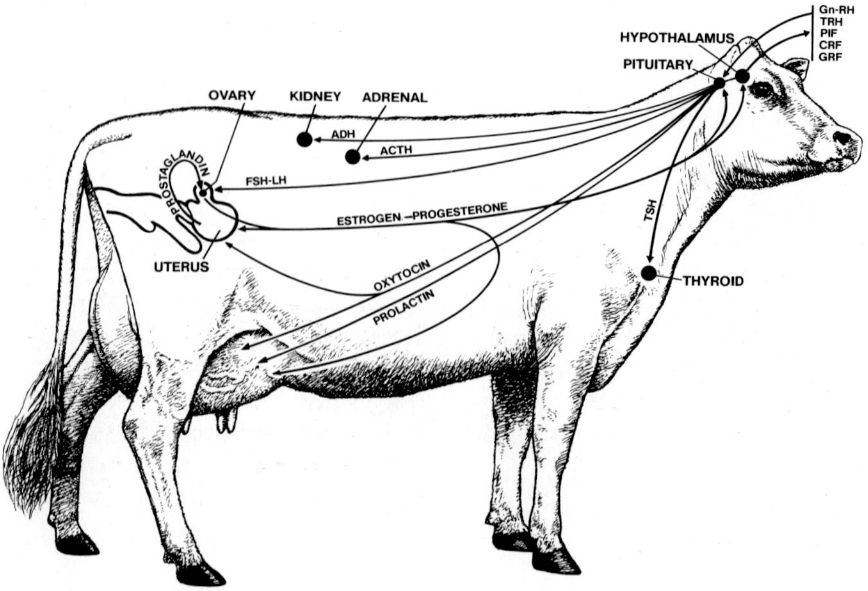

–Some may give lutalyse to all open over 45 days cows each Monday (you choose a day)

- when you see blood cow ovulated - 2 or 3 days past

Figure 2. On the left is an ovary with a follicle that contains the egg to be ovulated. It secretes estrogen, the hormone that causes the cow to exhibit the signs of heat. When the cow ovulates, the follicle ruptures and the egg moves into the oviduct. After ovulation, a corpus luteum (yellow body or CL) forms and secretes progesterone to prepare the uterus for pregnancy. (Courtesy of American Breeders Service)

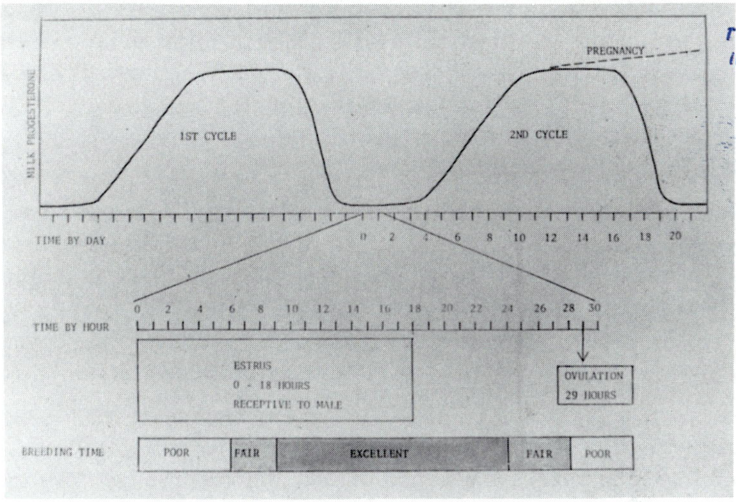

This hormone is secreted by the anterior pituitary gland at the base of the brain.

A condition related to LH is cystic ovaries. Cows with cystic ovaries develop follicles that fail to ovulate. These fluid-filled vesicles on the

Figure 3. A cow's progesterone level drops very suddenly when she comes in heat, so a low milk progesterone level immediately following a high level can be a good indicator that a cow is open and ready to breed. (Courtesy of IMC Pitman-Moore).

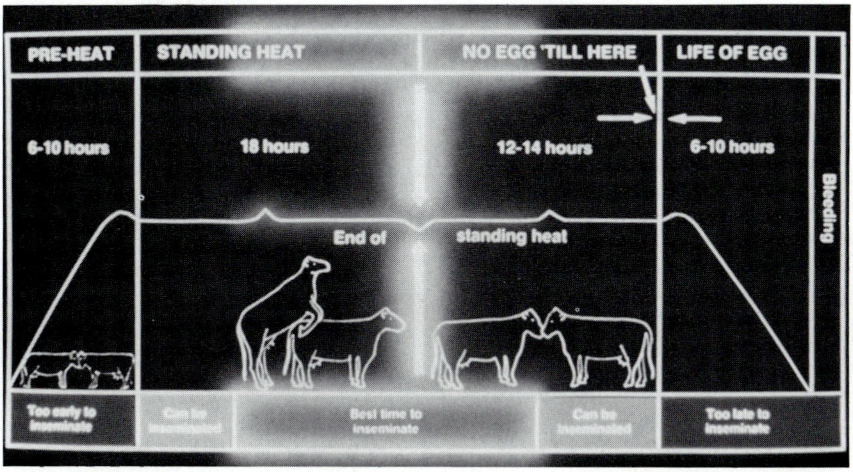

ovaries, called follicular cysts, cause some cows to show no heat cycles and other cows to be nymphomaniacs (constantly in heat). In still other cases, cysts occur because LH levels are inadequate to cause CL formation at the ovulation site.

Cystic ovaries may result from hormonal abnormalities primarily relating to LH and its interaction with other hormones. Low LH levels may cause follicular cysts or delayed ovulation. When ovulation occurs but luteinization (CL formation) is incomplete, this is referred to as a luteal cyst. Cysts may also develop with adequate LH levels but a lack of LH receptor sites on the ovaries.

Cystic ovaries develop for a number of reasons. Cows fed too little energy may become cystic. Diets low in the minerals needed to use energy cause a relative energy deficiency. Some of these key minerals are phosphorus, magnesium and cobalt. Some cows become cystic despite adequate energy and minerals. In these cases, the cow can't seem to eat enough to keep up with her high milk production.

Years ago, estradiol (ECP) in oil was commonly used to induce heat, and many high-producing cows became cystic from ECP therapy. This seldom happens today, but superovulation for embryo transfer in high producers may cause a very similar response.

There are many substitute LH products available to treat cows with cystic ovaries. Human chorionic gonadotropin (HCG), an extract from a pregnant women's urine, has LH-like action. Another is GnRH (Cystorelin), a synthetic compound identical to that produced by the hypothalamus. It has become very popular in the treatment of ovulation malfunction. Buserelin is a newly developed GnRH analog that has a longer action than GnRH. It shows promise of soon being available for routine use.

Prostaglandin

Prostaglandin (PGF_{2a}) is a hormone secreted by the uterine muscles (myometrium). It signals the ovary that the uterus does not contain a pregnancy. PGF_{2a} secretion dissolves the CL at day 17, after which progesterone levels drop rapidly. A new wave of follicular development then occurs in preparation for the next heat cycle. Because of this profound action on the CL, synthetic PGF_{2a} works well for inducing heat in cows not showing heat and cows being synchronized for embryo transfer. PGF_{2a} also causes uterine contraction and stimulates the uterine defense mechanism. These actions are useful when treating retained placenta. Controversy surrounds this type of retained placenta therapy and probably will not be resolved soon.

PGF_{2a} also can cause contractions of blood vessels. For this reason, people with asthma should handle it with care. One of my clients, Patty, is college educated and has a large beef herd. She used PGF_{2a} routinely to synchronize her best cows. Patty is also asthmatic. Once she accidentally spilled some PGF_{2a} on her hand and had an asthma attack from the skin contact. Patty gained a new respect for the drugs she uses on her cows. PGF_{2a} can also be used to induce abortion. For this reason, pregnant women should not handle it.

PGF_{2a} is available under many brand names, such as Lutalyse, Estrumate and Bovilene.

Follicle-Stimulating Hormone

Follicle-stimulating hormone (FSH) is secreted from the anterior pituitary gland and acts in conjunction with LH to cause growth, maturation and ovulation of follicles. FSH has 2 distinct surges during the estrous cycle. Between days 10 and 14, increased FSH levels cause the ovaries to grow many follicles during the luteal phase. The cow does not usually show heat after this surge. Later, at day 17, PGF_{2a} causes the CL to dissolve and another FSH surge occurs simultaneously. From this second surge, one follicle grows and matures for ovulation on day 0. Any other growing follicles shrivel up and disappear.

Vets and technicians routinely use FSH in embryo transfer techniques to cause a cow to ovulate many eggs (superovulation). FSH is available as FSH-P, an extract of sheep's brain. Pregnant mare serum gonadotropin (PMSG) also has FSH-like qualities. This extract of pregnant mare's blood acts like FSH.

Oxytocin

Oxytocin is a hormone secreted by the hypothalamus that causes milk letdown. It also causes uterine contractions and plays a role in stimulating uterine contractions during calving. Oxytocin also enhances the uterine defense mechanism and stimulates a cow's maternal instincts to mother her calf. Oxytocin is not normally a part of the cow's heat cycle. However, it will affect the CL if you give repeated doses hoping to stimulate milk letdown. You needn't be concerned with this effect if you only occasionally treat a cow with oxytocin.

Normal Heat Cycle

The interval of time between freshening and the first ovulation is short. In a normal, healthy, well-fed cow, it is 10-14 days. But various factors influence the timing. Ovulation is directly correlated to nutritional status. The cow must have received adequate protein and energy during her dry period for an early ovulation. She also must have gotten ample vitamin E, selenium, vitamin A, vitamin D, calcium and phosphorus, with optimum trace mineral supplementation.

The age of the cow also influences ovulation. Older cows tend to be slower in regaining reproductive functions after calving. Cows that have had an unsuccessful calving are also slow to cycle. Stresses such as from dystocia (difficult calving), retained placenta and milk fever may be the cause. The condition of a cow's uterus affects ovulation, too. A cow with septic metritis has a slow ovarian response.

Other extraneous sources of stress, such as hot weather, displaced abomasum and lameness, also delay normal cycling.

Cows with an energy deficit may grow follicles but not ovulate, and then develop cystic ovaries. Some cows may benefit from GnRH injections 10 days after calving. However, some cows develop pyometra (uterine infection) from GnRH injections after calving. Make the decision about using this therapy after talking with your vet. Also consider the cost of the hormone and improvements in the way you feed your dry cows.

The signs of ovulation are apparent if you're observant. They usually occur at the same time the uterus is clearing itself of tissue debris from the previous calving. First, look for reddish-brown material draining from the vulva. The discharge will be on the rear quarters of the cow and hanging from her tail. Because of the high PGF_{2a} and estrogen levels at ovulation, the uterine defense mechanism has a cleansing action. It evacuates any debris left from calving and sloughs the old endometrial lining off the caruncles. The underlying blood vessels also hemorrhage slightly. This is why the discharge looks somewhat bloody. At this time, the cow seldom exhibits signs of heat, such as riding other cows or standing to be ridden.

Don't think this kind of discharge at 10-14 days after calving means the cow has a bad uterus that needs infusing. In fact, infusing a normal cow can reduce the uterus' natural immunity. Here's the difference between a normal and abnormal discharge: an infected uterus produces a smelly, watery discharge. When the uterus is healthy, the discharge after calving is less watery and not foul smelling.

A cow's dietary energy level has a direct effect on ovarian and uterine health. Low dietary energy or low energy utilization because of mineral deficiencies may cause cows to have low LH levels, resulting in ovulation malfunctions. Weight loss because of high milk production creates an energy deficit. This ultimately results in lower LH levels and low conception rates.

Deficiencies of either or both vitamin E and selenium have been incriminated in retained placenta. Low dietary levels of these in the dry period contribute to cystic ovaries in mature cows. This deficiency also contributes to delayed first ovulations and prolonged days to first visible heat. Numerous reports indicate that dietary selenium should be supplemented at higher than legal levels. Erroneously, many veterinarians and dairymen think that an injection of vitamin E and selenium for springer cows is adequate in selenium-deficient areas (see Chapter 19).

All of these factors influence your profits. Factors delaying the onset of cycling delay the first heat, cause a longer calving interval, and reduce profits every day the cow is open more than 100 days.

For maximum profits, every cow must have a calf every year. If the calving interval is more than 12.5 months, production is not at its peak. Profits are highest when cows calve as often as possible. The more pregnancies and lactations, the more profits. To reach her maximum production potential, a cow should be pregnant by 100 days after freshening. Heat must be detected early in lactation to get the cow pregnant by 100 days. This is because it usually takes 2 services for conception.

How to Make Heat Detection Pay Off

For most dairymen, heat detection is a chore done with little enthusiasm. Certainly, it is time consuming and has little immediate reward. However, it has far-reaching financial consequences when done poorly.

More cows cycle in the evenings and at night than during the middle of the day. Cattle should be observed for signs of heat very early in the morning and after the milking is done in the evening. If you milk 3 times a day, observe your cows as late as 10 PM. If the cows are in a dry lot, observe them out on the lot where the footing is good. Watch the cows for at least 20 minutes. A cow may mount another cow or stand to be mounted only about once every 20 minutes. This is the reason it's important that you watch your cows for at least 20-30 minutes. The cattle should be observed by the person working around the cows on the lot. He or she should make a conscious effort to watch for signs of heat, rather than rushing cows into the holding pen or running the skid loader.

Most farmers think an easy time to watch for heat detection is while the silo unloader is running. This is not a good time because if a cow is hungry, she will spend her time at the bunk instead of riding another cow.

Figure 4. A cow ovulates after she goes out of heat, so the best time to breed a cow is near the end of standing heat. (Courtesy of American Breeders Service)

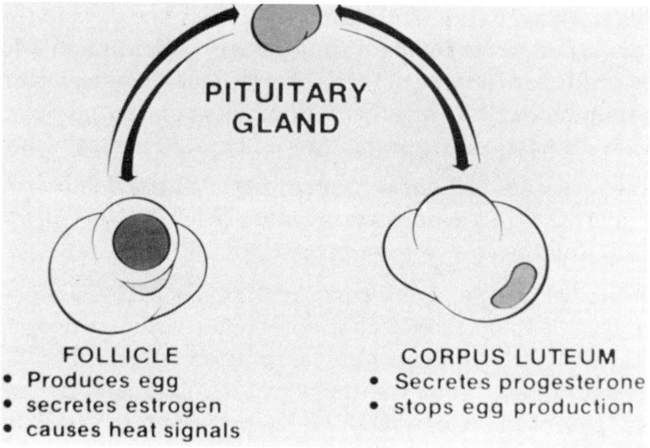

Signs of Heat

A cow standing firm to be mounted by another cow is the most consistent sign of heat (Fig 4). Other signs of heat include a mucus discharge from the vulva, swelling and redness of the vulva, bellowing, restlessness and riding other cows. Some cows in heat only stand a short time and spend the rest of their time riding other cows. Cows riding other cows should be checked to see if they are coming into heat or going out.

Heat can be detected in various ways. Most important, somebody must watch the cows. An aid that can be used to augment observation

Figure 5. Cows in heat can be identified by a bull with a chin ball marker. When the bull rests his chin on or rides a cow in heat, the chin ball marker releases red paint, like ink from a ballpoint pen. (Courtesy of American Breeders Service)

is K-Mar Heat Detectors. With this system, a patch is glued on the cow's tailhead. The heat detector patch turns from white to red when a cow mounts and she's standing firm. A little capsule of red dye in the patch is ruptured when the other cows ride her. A second aid is heat detection paint. This works like the heat detectors. The paint is put on the cow's tailhead. The paint is rubbed off when another cow rides her.

Another method is a marker bull (Fig 5). This is a bull that has been surgically altered so that he cannot breed a cow. The bull has a ball marker strapped under his chin so that when he mounts a cow or rests his chin over the cow's back, he paints the cow. A possible disadvantage is that the marker bull may "fall in love" with one cow. He may spend all his time mounting one cow and marking her, rather than looking for other cows coming into heat. Therefore, it's important that the cow in heat be removed from where the bull is working. When she's gone, he'll spend his time seeking other cows.

A marker bull can be surgically altered by a number of different procedures, but one method that is *not* recommended is a vasectomy. This allows the bull to penetrate the cow's vagina and spread disease.

There are other gadgets available for heat detection but most aren't very practical. A probe device can measure the electroconductivity of the vaginal mucus. A pedometer strapped to a cow's leg indicates how many miles a cow has walked. Generally, a cow in heat does a lot more walking. A number of biotelemetry devices are being developed to aid in heat detection.

Anticipating a cow's heat is an important aid to heat detection. On a regularly scheduled reproductive health visit, your vet can palpate the cows and determine which have cycled and have CLs. These cows are given PGF_{2a} to bring them into heat. The herdsman can then estimate approximately when these cows will be in heat and be watching for them. Having several cows cycling within 1-2 days of each other often improves heat detection for the whole herd. In addition, the PGF_{2a} helps clear up any residual metritis (uterine infection).

Another method of heat detection is an ear implant called Synchromate B. This has not been approved for dairy cattle and should only be used under the direct supervision of your vet.

Breeding: Natural Service or AI?

Once heat has been detected, the cow must be bred by artificial insemination (AI) or natural service with a bull. There are advantages and disadvantages to each.

Natural Service

Natural service has several advantages. Most important, the bull does all the heat detection. He has an incentive program that makes him a tireless worker.

On the down side, here are some of my observations. Generally, producers that use strictly natural services do not have the management expertise of top producers. It's the rare herd that uses all natural service and has top production.

Also, there's a real danger to the dairyman. Every year dairymen are killed or maimed by bulls. Furthermore, the incidence of venereal disease increases when a bull is used in the herd. Genetics are seldom improved by using a bull. And there are additional costs involved in maintaining the bull, such as fences and pens.

Such is Life

One time I stopped to treat a sick cow at Joe D's. He had his bull hitched to a yoke and harness. After we finished with the sick cow, Joe hooked the bull to a single bottom plow and proceeded to the field. When I asked what he was doing, Joe spit his tobacco out and replied he "was just teaching his bull that there was more to farming than breeding cows and tearing up fences."

Artificial Insemination

Artificial insemination has some advantages. It allows you to select the best bulls to match the traits of your cows. You can select bulls that sire for easy calving, which is especially helpful in heifers. You can select for bulls that will improve the rear udder suspension of the offspring. By evaluating the statistical performance report of the bulls used in a number of herds across the country, you can pick those with superior improvement in the bull's daughters versus their herdmates. This is called predicted difference and repeatability. You also can pick bulls that sire daughters with good feet and legs.

When using AI, you are not limited to one bull. A different bull can be used on each cow. AI bulls have been measured statistically for their butterfat production and protein production. Bulls can be continually selected for high butterfat and protein. These traits are highly heritable, so your herd can show rapid genetic improvement.

With AI, the spread of venereal disease can be prevented at mating. You can economically choose which bulls will improve the traits of your herd. Top-of-the-line genes are available to anybody in the country, regardless of economic situation. It also gives any dairyman the ability to sell breeding stock based on the reputation of the sires that have been used in his herd.

There are some disadvantages, however. First, it requires extra expertise on your part. Artificial insemination also requires more time than natural service. It takes several minutes to breed a cow when maybe you should be planting corn, milking cows or making hay. You must do all the heat detection. Unfortunately, dairymen don't have the same incentive to detect heat that the bull does. Dairymen who do not have large herds may have a difficult time maintaining their AI expertise, so a technician is needed. Managers with large herds are better at maintaining this level of expertise to inseminate their cows.

Let's assume you choose AI. Selecting bulls and purchasing frozen semen are important for the future of the herd. When you use semen from a particular bull, *you're making a decision that will affect the economics of your herd 6 years into the future*. Here's why: If she conceives, the cow will be pregnant 9 months. Then the heifer calf born from this pregnancy will grow for 2 years before she starts making a positive economic contribution to the herd. She'll stick around as a producing cow for about 3 years before you cull her. You get the idea.

I urge you to buy semen from a reputable bull stud. Don't buy from "semen jockeys" that come in off the road, because you can't trust the quality of their product even though their intentions may be good. You need assurance that the frozen semen has been handled correctly. When that's not the case, no amount of skill on the part of your AI technician will make up the difference. You'll end up with poor conception rates or, worse yet, no pregnancies at all.

Handling Semen

The temperature of frozen semen in the tank must not rise to freezing. Keeping semen frozen is not enough. It must be maintained at the temperature of liquid nitrogen, which is -190 F. When using semen in straws, make sure each straw is in a plastic goblet in a rack. This maintains the correct temperature during transfer of the racks of semen from one tank to another. If the straws are not in small amounts of liquid nitrogen in these goblets, the temperature of the straws rapidly rises and damages the sperm.

When transferring frozen semen from one tank to the other, check first to see that there's nitrogen in both. When you move a straw from one tank to another, move the whole goblet, and do so very quickly (Fig 6). There's a rapid heat exchange if the straws are exposed to air very long. The smaller the volume of the straw, the faster you must move.

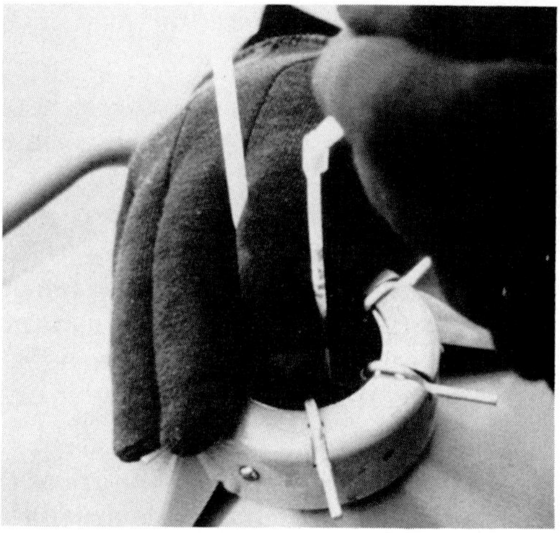

Figure 6. You must be careful to prevent rapid temperature changes in a liquid nitrogen tank when checking the semen tank inventory and removing semen straws. Semen is easily damaged by rapid changes of temperature in the neck of the tank. The temperature of the semen rack rises rapidly when held in the neck of the tank for more than a few seconds. (Courtesy of Dr. D.A. Morrow and W.B. Saunders. From *Current Therapy in Theriogenology*)

Aim for an elapsed time of 10 seconds or less. Even though it's been reported that one cow conceived from bull sperm that had been stored for 19 years, keep your herd out of this kind of record book. Avoid buying frozen semen far ahead of the time you'll use it. Try to keep your inventory low so you're always using fresh frozen semen to breed cows.

When transferring semen from the tank to the thaw box, do so quickly. Bring the rack out of the canister into the neck for only a few seconds, because it's warmer there than down in the canister. And always keep the canister at the neck level or lower (Fig 7). Lower the canister back into the tank after you've made each semen selection so the temperature in the canister has a chance to stabilize.

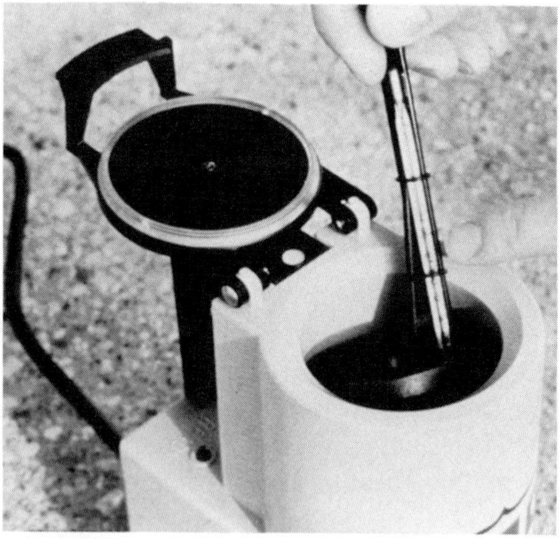

Figure 7. The rack of semen straws contains a goblet that keeps each straw bathed in liquid nitrogen. When removing semen, do not bring the goblet up higher than the neck of the tank. To avoid frostbite, wear gloves when handling frozen semen. (Courtesy of American Breeders Service)

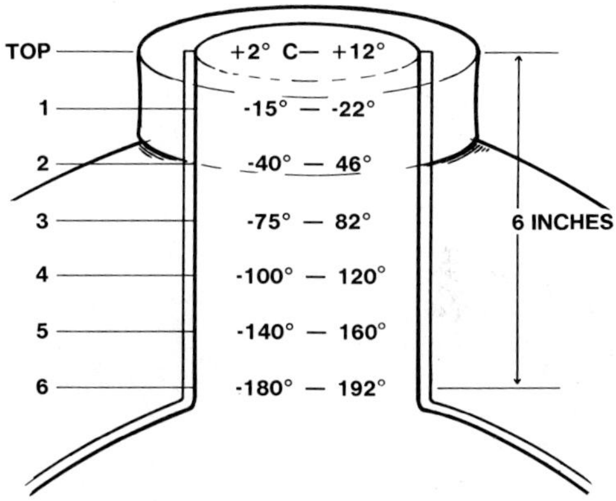

Figure 8. Semen must be thawed at the correct temperature to have live sperm for conception. Consult your semen supplier for recommendations regarding correct thawing temperatures. (Courtesy of American Breeders Service)

The same rules apply to ampules. Transfer them in 10 seconds or less. In general, ampules aren't as vulnerable to cold shock as straws. But for best results, handle them as though they were.

Keep an inventory sheet current on your tank. This reduces the search time for individual straws and ampules when you have the canister raised up in the neck.

Thawing Frozen Semen

Here are the procedures for thawing frozen semen (Fig 8). With ampules, certain basic equipment is required. First, you'll need a thaw box with either ice water or 95 F water (depending upon bull stud recommendations). Locate a dial thermometer so you can monitor this temperature. Use forceps to remove the straw or ampule from the rack

Figure 9. Use forceps to lift the straw of semen from the goblet, and then drop the straw immediately into the thaw box. (Courtesy of American Breeders Service)

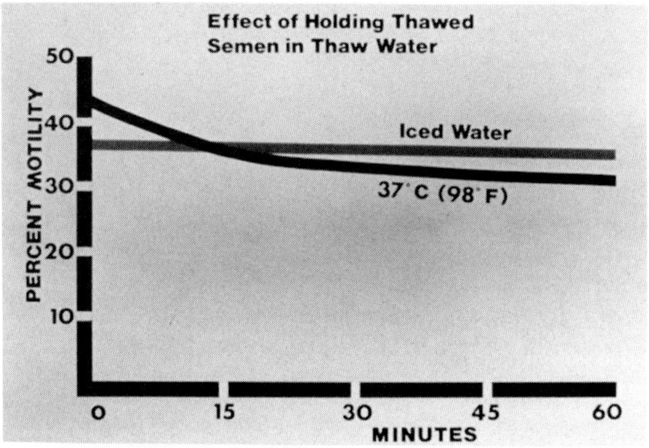

Figure 10. Sperm motility is directly related to thawing time. Semen soon loses its motility if it is left in the thaw box longer than recommended. (Courtesy of American Breeders Service)

and drop it into the thaw box (Fig 9). Gloves will protect you from frostbite when handling the rack in the canister. A scribe allows you to score the ampule so you can break open the top to remove the semen. Follow this sequence:

- Open the tank.
- Identify the canister containing the ampule you want.
- Raise the canister up to the neck.
- Raise the rack up high enough to remove the ampule.
- Quickly remove the ampule and drop it in the thawing water.

Be careful to protect your face from the thaw box when you drop the ampule in the water. If the semen was not properly processed at the bull stud, the faulty ampule may explode, sending broken glass into your face. There's no experience quite like dropping an ampule in a thaw box and having it blow up, sending glass shards and semen all over the ceiling! It usually takes a minute for your blood pressure to drop back to normal.

Leave the ampule in the thaw box for at least 10 minutes. Do not break the ice off the ampule before the semen is thawed. Instead of a gradual thaw, the sudden temperature rise will damage the sperm. Once the semen is thawed, peel the ice off the ampule, scribe or score it, and with a paper towel covering your fingers, break the top. Withdraw the contents of the ampule into a prepared breeding pipette. Your goal is to end up with a solid column of semen in the inseminating rod.

Straws are handled differently than ampules. You'll need these supplies: a thaw box complete with a thermometer and a pint of water; scissors to open the straw; and an AI gun (or Cassou gun) to load the semen. Follow this procedure:

- Regulate the water temperature to 95 F.
- Identify the canister you want.
- Raise it into the neck of the tank.
- Raise the rack from the canister.
- Use forceps to remove the straw from the canister.
- Lower the rack with one hand, and use the other to place the straw in the water bath.

Thaw the straw for 30 seconds or the period recommended by the stud. Time is critical, so don't thaw it longer than necessary (Fig 10). Wipe the water off the straw. The semen must not be exposed to water because it may damage the sperm. Clip the sealed end of the straw with the scissors at a 90-degree (right) angle to the straw (Fig 11).

Place the straw with the cotton-plugged end first in the Cassou gun. This you'll cover with a plastic protective sheath, and a plastic retainer washer will hold the sheath in place. Next put a protective double sheath over the AI sheath on the gun (Figs 12, 13). This is called a double rod and it prevents transmission of venereal disease into the cow's uterus from her vagina.

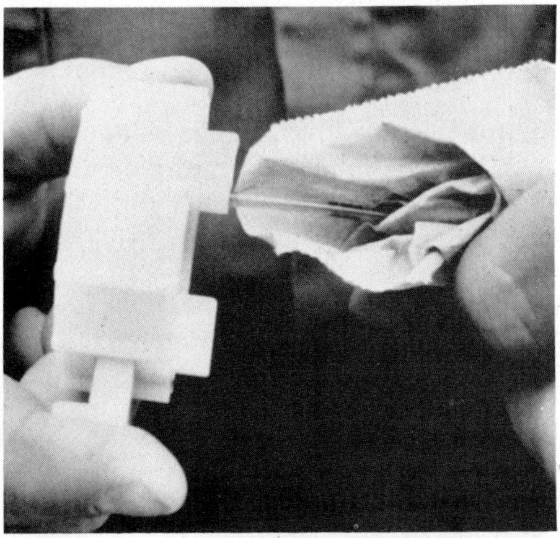

Figure 11. After the straw of semen is thawed for the recommended time, wipe the water off the straw and cut the sealed end squarely for a snug fit in the sheath. (Courtesy of American Breeders Service)

Regardless of whether you use straws or ampules, heed these precautions. Once the semen is thawed, avoid exposing it to direct sunlight. Also, avoid exposing thawed semen to cold temperatures. And *never* prepare the semen until the cow is restrained! The sperm may begin to die while you're chasing the cow around.

Figure 12. With an O-ring syringe, insert the straw into the end of the syringe before sliding on the sheath. (Courtesy of American Breeders Service)

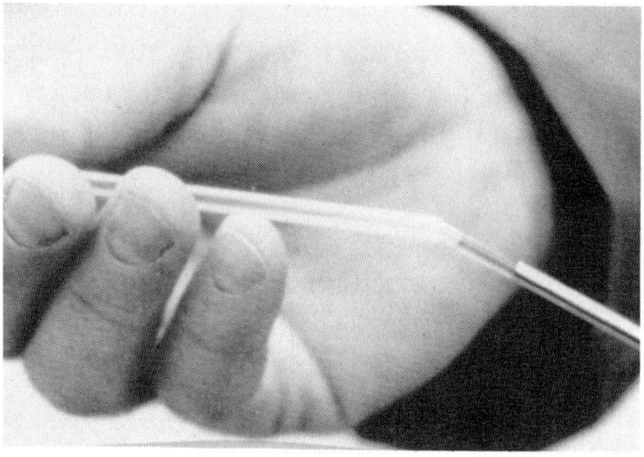

Inseminating the Cow

Next comes the actual insemination. You may think that once you've talked to the veterinarian or technician and know what a uterus or cervix looks like, you're ready to breed cows successfully. But there's no substitute for proper training in AI techniques. You need to develop special skill in threading the rod past the cervix into the uterus.

Time the breeding so it occurs at or near the end of the standing heat. Clean away any manure from around the tail area (Fig 14). Throughout the procedure, handle the cow with care, minimizing any stress. Excitement reduces conception rates. When a cow is nervous because she's

Figure 13. The O-ring adaptor is at the top of the syringe. Note the snug fit that is achieved when all is in place. (Courtesy of American Breeders Service)

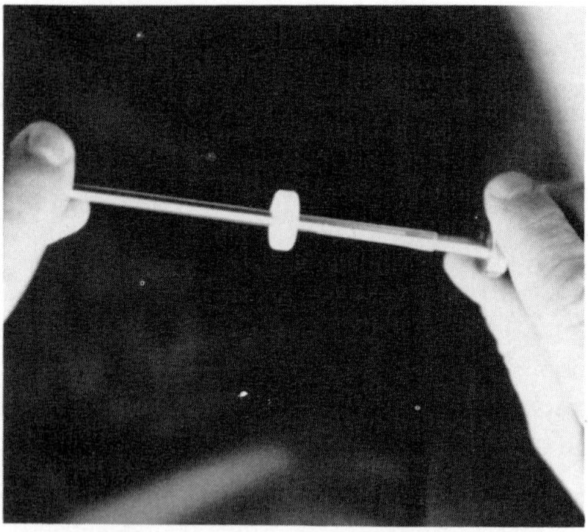

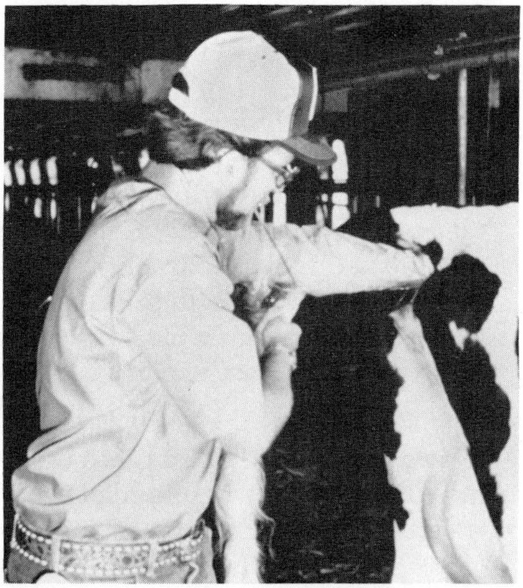

Figure 14. Before inseminating, wipe the manure and extraneous dirt from the vulva. (Courtesy of American Breeders Service)

been chased all over the pasture before breeding, her adrenaline levels will be high. This antagonizes the positive effect natural oxytocin has on the uterus in transporting the semen to the oviduct.

Follow these steps to inseminate a cow.

- Load the pipette with semen and hold it in your teeth while preparing the cow (Fig 15).
- Lubricate your gloved hand and put it into the cow's rectum (use your less-coordinated hand for this job).
- Clear the rectum of manure.
- Wipe the vulva clean with paper towels.
- Locate the cervix.

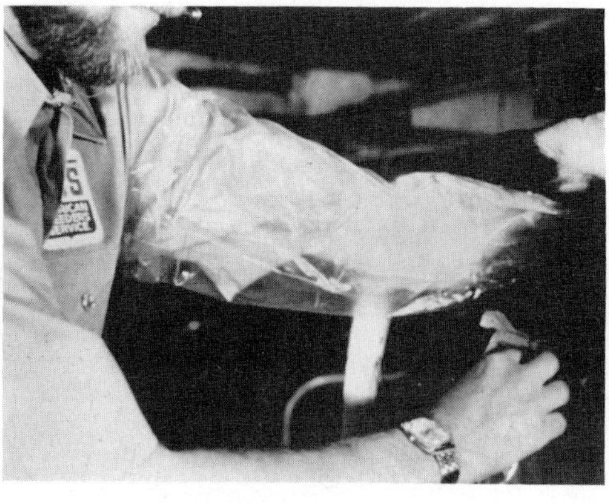

Figure 15. While holding the straw gun in his teeth, the inseminator uses his sleeved hand to manipulate the cervix rectally. (Courtesy of American Breeders Service)

- Insert the inseminating pipette into the vagina until you reach the cervix.
- Push the Cassou gun (with breeding sheath) through the end of the double rod.
- Push the cervix forward a bit with your hand in the rectum to take the wrinkles out of the vaginal wall (do this gently — any blood on the rod tip may kill sperm) (Figs 16, 17).
- Thread the pipette through the cervix.
- Deposit sperm in the body of the uterus just beyond the cervix as you withdraw the pipette.
- Work to make the whole procedure a rapid, coordinated event.

Never thread the pipette into the uterine horns or attempt to direct semen to the side where ovulation is going to occur. This maneuver lowers the conception rate (Figs 18-20).

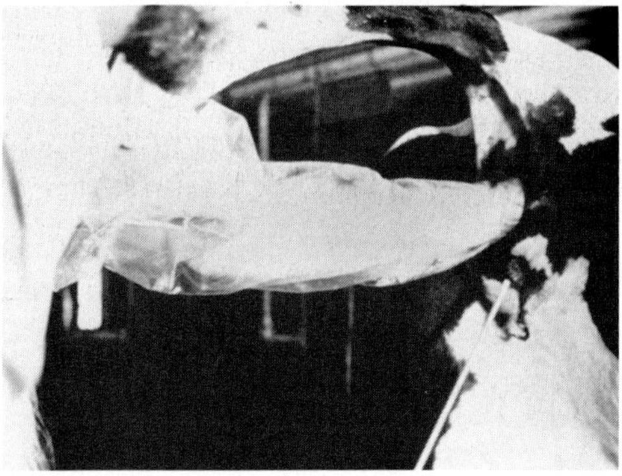

Figure 16. Applying downward pressure with the gloved hand in the rectum opens the vulva so the straw gun can be inserted into the vagina. (Courtesy of American Breeders Service)

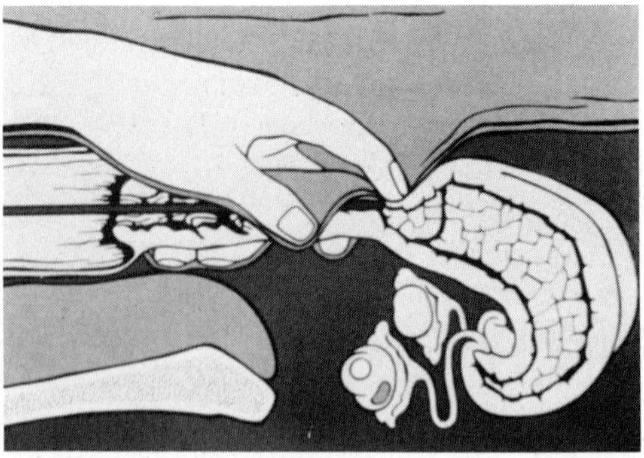

Figure 17. The gloved hand in the rectum holds the cervix, while the other hand is used to thread the straw gun through the vagina and into the cervix for deposition of the semen on the anterior side of the cervix. Note that the uterus and ovaries are coiled tightly just in front of the hand on the cervix.

Figures 18 (top), 19 (left) and 20 (right). Veterinarians use radiographs (x-rays) to show when semen has been deposited correctly or incorrectly. The semen must be deposited just through the cervix. In which of these photos was the semen deposited correctly? (Courtesy of American Breeders Service)

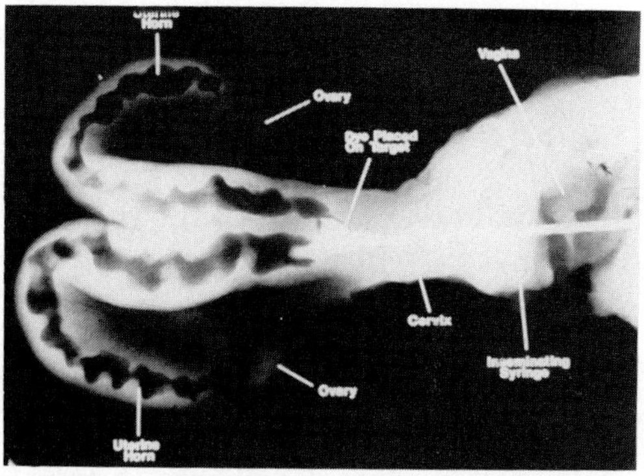

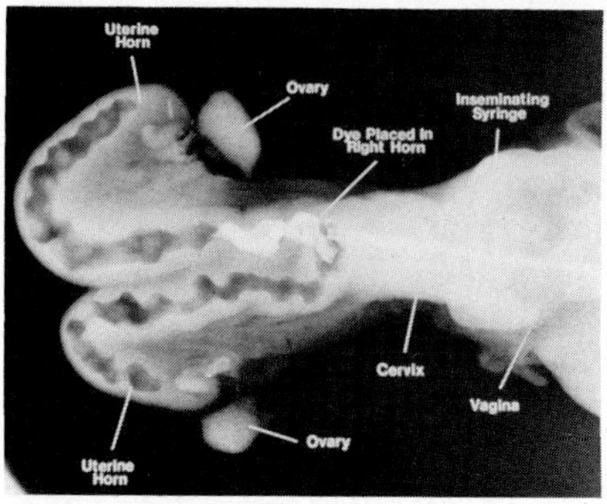

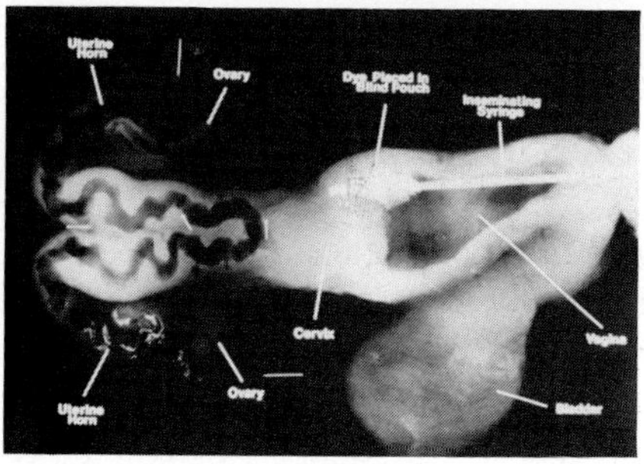

(Answer: Figure 18)

When to Use Embryo Transfer

You may want to integrate some embryo transfer into your breeding program. This technique transfers genetic material from a superior dam to a grade dam that carries the calf to term. This lets you get the most out of your superior dam by increasing the number of offspring she'll produce annually. Or you can propagate a cow's good genes even if she can't maintain her own pregnancy. This technique boosts the economic importance of the donor cow.

Pros and Cons

Quite a few cows with superior genetics have infertility problems. Embryo transfer bypasses the infertility of a hard-breeding cow. It may even restore her fertility, as it allows cleansing of the uterine lining to remove old tissue debris. Embryo transfer may also rejuvenate malfunctioning hormonal systems in an older cow.

Bulls are part of the picture, too. You can evaluate bulls for genetic defects by superovulating a known carrier of a genetic defect and breeding her to a bull with the suspected recessive trait. Then you put the resulting embryos in a number of recipient heifers. In just 60-90 days, you can remove the fetuses by cesarean section and evaluate them to determine if they have the genetic defect. A good example of this procedure is in diagnosis of mule foot. Embryo transfer is also a good way to evaluate the fertility of bulls.

There are some drawbacks to this procedure. First, it's expensive. Another disadvantage is that genetic gain on the female is substantially less than the genetic gain of selecting good bulls.

Selecting Donor Cows

If you try the technique, there are criteria for selecting a donor cow. First, determine if you can profitably sell bull calves that result (if you're flushing to get heifers; the opposite is true for marketing heifer calves if it's bull calves you're trying to produce). You want to recover the cost of the transfer for all of the calves. Next, choose a cow that's in the top 2% of her breed. It does little good to do a transfer on a cow that offers ordinary genetics. Avoid doing embryo transfers for the wrong reasons. I've heard dairymen say they chose a donor cow because she won grand champion at the State Fair, or because her half sister sold for a good price. I've even come across the occasional dairyman who did embryo transfers because he had a lot of open heifers he wanted pregnant!

Who Will Do It?

You might need some help selecting a veterinarian or an embryo transfer company to do the procedure. Their technical staff should be highly regarded, and their pregnancy rates high. It's important to know if they do the transfers surgically or nonsurgically. If they're doing a surgical procedure, the pregnancy rate should be consistently in 70-80% level on the eggs transferred. A good range for nonsurgical procedure is 50% or higher.

Though I seem to be emphasizing pregnancy rates, there's more to the story. These figures are also influenced by the quality of embryos implanted. If a company puts in the very best embryos they recover, their pregnancy rates will be better than one that transfers every single fertilized embryo. The pregnancy rate may be lower for the company that puts in all of the embryos, but they ultimately salvage more embryos for the dairyman.

It's also a good idea to inquire about whether embryo freezing is available. It's frustrating to end up with a donor that ovulates 15 embryos when you only have 10 recipients ready. If freezing is available, you can save those extra embryos until you have more recipients available. No sense wasting good embryos for lack of recipients. Last, check the reputation of the embryo transfer company or vet with other dairymen. They'll tell you from their experiences if they've been satisfied with the services offered.

Getting Ready

The first step in an embryo transfer is to chart the donor cow's heat cycles so you know she's on a regular 21-day schedule. If you can do the same with your recipients, you can try to match their cycles to the donor. The goal is to synchronize them all so that the donor and the recipients all come into heat on the same day.

If things match up, you can superovulate the donor with FSH to cause ovulation of more than one egg. But if you need to get the recipients synchronized first, they should receive an injection of PGF_{2a}. A week or so later, start the donor's superovulation schedule. The FSH works best during the luteal phase of the donor's cycle (between days 6 and 16). Usually the hormone is injected twice daily for 4 days. Before the last day's injections, the donor is injected with PGF_{2a} to bring her into heat. Do the same for the recipients one day ahead of the donor. This is because recipient heifers usually come in heat in about 72 hours, while the donor responds in about 48 hours.

Some donor cows with a history of cystic ovaries or other ovarian dysfunction benefit from therapy with progesterone implants. Make sure you follow the advice of the veterinarian doing the embryo transfer.

Good nutrition is a stepping stone to the success of any embryo transfer program. This goes for the recipients as well as the donor cow. You may need to give the donor supplements in addition to her milking ration. Thin recipients need more energy and protein so they can gain some weight. It's a good idea to get them started on a better ration at least 30 days before you plan a transfer. Also pay attention to optimum levels of vitamin E, selenium, copper and other trace minerals.

Assuming the superovulation and synchronization work, the donor and the recipient cow will come into heat on the same day. Watch carefully for heats and keep detailed records. If the donor and recipients are off by 12 hours, you're not in trouble. But if the cycles are more than 24 hours apart, results will be poor.

Breed the donor cow with good-quality semen that has been previously checked for fertility. Then breed her 2-3 more times during her heat period: in the early stages, in the middle, and after she's gone out of heat. Superovulation causes her to ovulate a number of eggs throughout the heat period. She needs extra sperm to fertilize the increased number of eggs.

Some breeders use semen from more than one bull on the cow. But you must plan ahead so you can prove parentage of offspring. Your breed organization must verify each bull's blood type before breeding. For example, it's possible to end up with a total of 5 calves: 2 from one bull and 3 from another bull. If the calves' blood types are different, the sires can easily be identified. If not, you have no way to document parentage.

After the breedings, some embryo transfer vets use LH or GnRH on donor cows to ensure ovulation. Then 6-9 days later, you set up the actual embryo transfer.

Equipment

Doing it at an embryo transfer company facility is pretty simple. You take the donor there and the technicians go to work. Embryo transfers done on the farm work as well as those done at a special facility. But you'll need to rig up some basic equipment.

Set up a chute, stanchion or headgate to restrain the donor cow. Place it so the cow's front quarters are higher than the back quarters. In a tie-stall barn, you can do this easily by backing the cow up, placing a 2 x 4 across the front of the stanchion so the cow must stand with her hindlegs in the gutter. This provides enough front-end elevation for the flush procedure to work effectively. Other dairymen have the cow stand on a wooden box so she faces "uphill" (Fig 21).

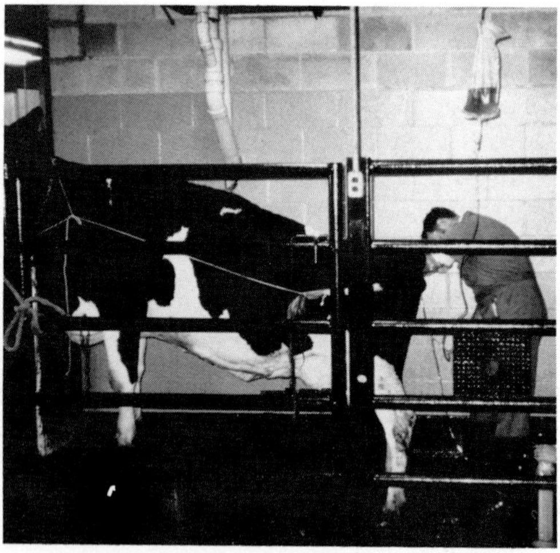

Figure 21. A catheter irrigates the uterus with sterile fluid for embryo recovery. The cow's forequarters are elevated to help the fluid drain from the uterus.

Clean up your facilities, remove cobwebs, and control flies. Provide a table for the vet to use so his instruments and solutions are close at hand. Since embryo transfer is done by irrigating the uterus with a fluid container suspended over the cow, you'll need a way to hang an IV bottle overhead. Think about how you're going to restrain the recipients, too. They're going to receive spinal anesthesia so that implantation can be done quickly and safely. The recipients will have an egg placed in them either by surgical or nonsurgical means.

The Procedure

After the donor cow receives spinal anesthesia, the vet inserts the flush system through the cervix into her uterus. It goes in only one uterine horn at a time. Then he removes the catheter, and places another in the opposite horn. Occasionally, one catheter is used to flush both horns.

The ova have been fertilized and are now embryos that have traveled through the fallopian tubes into the uterine horn. Flushing brings the embryos out. The flush solution contains special proteins and antibiotics designed to keep the embryos alive. The vet fills and drains the horn several times. The dislodged embryos are captured on an embryo filter in a receptacle, or the solution is drained into a cylinder and the embryos settle to the bottom. If the embryos were captured on a filter, they are washed into a Petri dish for examination. If they're allowed to float into a cylinder and settle to the bottom, the upper fluid is siphoned off and the embryos are left in a small amount of remaining fluid.

Next, the vet uses a dissecting microscope to search the fluid for the embryos. This may take a few minutes or up to an hour. The embryos are only as large as a particle of dust, so you can't see them readily with the naked eye. Each embryo is evaluated for health and viability (Fig 22). The vet will cull unfertilized eggs and embryos that are immature, defective, deformed or dying (Fig 23). The embryos are then put through a number of wash solutions to remove any disease contamination.

From the number of viable embryos recovered, you'll know exactly how many recipients you'll need. Surgical transfer requires that the recipient be clipped on the side where ovulation occurred, surgically scrubbed and then given a local anesthetic. The surgeon makes an incision and works his way into the abdominal cavity. He pulls the uterine horn to the incision site, and puts a small needle hole in it. A technician picks up an embryo in a plastic catheter and places it into the uterine horn via the needle hole. At this point, the vet replaces the uterus to its normal position, sutures the incision, and the recipient recovers nicely.

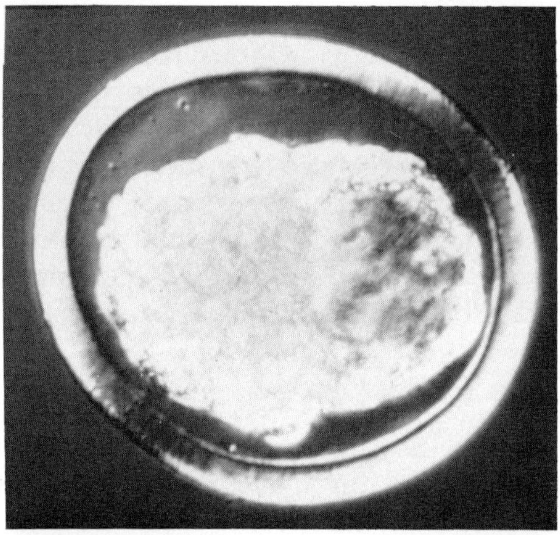

Figure 22. This 8-day-old embryo is ready to be implanted into a recipient heifer. In 2-3 more days, the embryo begins to develop recognizable organs and body parts.

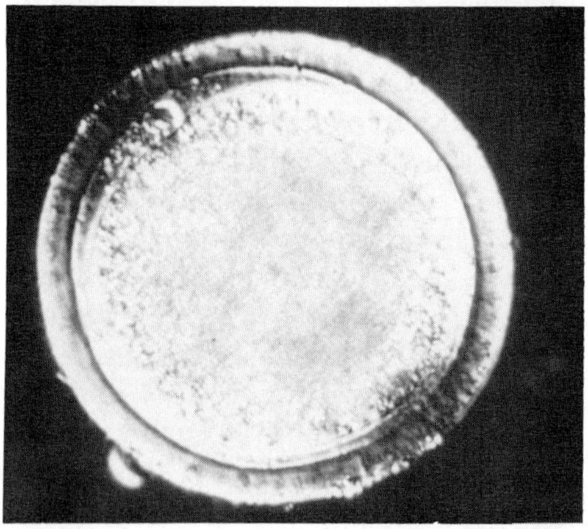

Figure 23. Unfertilized eggs are often recovered in the flush from a donor cow. Shown is a degenerating, unfertilized egg.

With the nonsurgical method, each recipient is given a spinal anesthetic so she does not strain during the procedure. The embryo is loaded into an AI straw and a Cassou gun. The vet puts a gloved hand in the rectum, cleans the vagina with a paper towel and threads the loaded Cassou gun with a double rod through to the cervix. The Cassou gun is put through the end of the double rod and, instead of depositing the embryo just through the cervix as in AI, he gently threads the rod up the uterine horn as far forward into the horn as possible. This must be done skillfully to avoid hemorrhage, which can kill the embryo. After depositing the embryo, the Cassou gun is withdrawn. Recovery is usually uneventful.

There are advantages and disadvantages to each of these procedures. Surgical transfer has a higher pregnancy rate. Also, when the incision is made, if there is something wrong with the CL or the uterus that wasn't detected on palpation, it gives the veterinarian the opportunity to inspect and stop the transfer. The disadvantage to surgical embryo transfer is that it involves more work and major surgery. While most embryo transfer surgeries are uneventful, there is some risk of infection afterward.

Nonsurgical transfer is easy for an experienced technician and it's quick and inexpensive. The disadvantage of nonsurgical embryo transfer is that the pregnancy rate often is 20-30% lower than for surgical transfer. If the embryos are valuable, some are sacrificed using the easier, quick-transfer method. The person doing nonsurgical transfer requires a good deal of practice and skill to be successful.

Embryo freezing is a good alternative when not enough recipients are available for the transfer. It also cuts the cost of boarding a number of recipient heifers while waiting for a transfer. If you plan to freeze embryos, the vet flushes 1-2 days earlier than usual. Freezing the embryos at this earlier stage of development improves their survival at thawing. Generally, the pregnancy rate with frozen embryos is about half of the pregnancy rate with surgically implanted embryos.

Sexing procedures can determine the sex of embryos. A reliable means of sexing is by chromosomal assay. Unfortunately, the embryo death rate from this procedure is high because the assay is traumatic to embryos. New sexing procedures are being developed but are not yet commercially available.

Followup Care

Flushing a superovulated cow stretches the uterus out of shape, and it takes time for it to recover. Generally, I recommend about 30 days' recovery time. It's also wise, immediately after the flush, to give PGF_{2a} to the donor cow. This hastens the removal of the many CLs that may persist on the ovaries after superovulation.

11

Managing Repeat Breeders

DHIA records show that the average calving interval is about 13.5 months. This is 54 days longer than the economically ideal calving interval of 12 months. Long calving intervals cost you about $2-3 per day in lost profits from high production in early lactation. This amounts to an annual loss in excess of $900 million to all dairymen. The losses from culling plus the losses from excessive calving intervals total about $180 per cow per year from infertility in everyone's herd.

What are Repeat Breeders?

A cow should be labeled a problem breeder or repeat breeder if she has been bred 3 times without becoming pregnant. Many dairymen breed cows every 21 days and don't become concerned until a cow has been serviced 5-6 times. You should have the same concern for cows that have not been seen in heat by 60 days after calving or have been diagnosed open after breeding as you have for the long-term repeat breeder.

If cows are not confirmed pregnant by 100 days after calving, it is not possible to maintain a 12-month calving interval. To run a profitable operation, you must do everything possible to prevent a cow from becoming a problem breeder. Profits are highest when a cow has a calf every 12 months. She must be pregnant by 100 days after calving if she is to even come close.

Repeat breeders are a financial loss to dairymen, especially when a high-producing cow is sold because of infertility. The dairyman not only loses a valuable member of the herd, but also loses the opportunity to cull a less valuable, lower-producing herdmate. In most geographic areas, involuntary culling accounts for up to 30% of herd losses each year. Infertility is the most common reason for involuntary culling. A conservative estimate of the average difference between salvage price of the cull cow and her replacement is $400. This loss amounts to about $360 million for all dairymen annually.

Repeat breeding may be related to the bulls used in natural service or related to the technicians doing artificial insemination.

A Case in Point

Ben is a taciturn old farmer who has little to say about his herd problem, other than he has too many repeat breeders. He has a picturesque farm with all of the barns and silos neatly painted white. On my first visit, he volunteered mostly a "yup" or "nope" to questions about his cows. In between, he spat tobacco juice on the concrete. He has chewed tobacco enough years that he has a permanent pouch in his cheek where he keeps the chaw.

On my regular herd visit, Ben usually brought in "problem" cows to examine, which consisted of most of the herd. His Holsteins were all quite fat, many having big patches of fat on their hips and pins.

Ben is one of my dairymen who had just "too many repeat breeders." Ben's records were also in bad condition. He took notes on assorted cow towels and later transposed the notes into a notebook. He couldn't read the vet's writing so he wasn't sure what some of his records said.

One of Ben's basic problems was that he didn't recognize a potential problem cow developing until she was already well into her lactation. His records were just not adequate. In most cases, he didn't recognize a cow as a potential problem until she was diagnosed as open on palpation. He didn't see his cows in heat 21 days after breeding, so he had to rely on the vet's visit. Involuntary culling was Ben's chief complaint. It was his best cows that didn't breed, not his loafers.

Infertility is a major reason for involuntary culling, not only in Ben's herd but in most dairy herds. When valuable cows must be culled or have a long lactation, genetic progress is slowed. The average milk yield for the herd and profits are reduced with fewer cows freshening.

Causes of Infertility

Reproductive failures are all too common in today's herds. There are many misconceptions (no pun intended) about infertility. Most important, it is *not* a disease. This was contrary to Ben's opinion. All he wanted was a shot to make his cows come into heat, ovulate and get pregnant. It took several visits to convince Ben that infertility is not a disease that a shot could cure. Infertility may be a result of many different factors. Sometimes it stems from a malfunction that traces back to calving, or from how and what cows are fed.

Each herd must be diagnosed and treated according to the causes and types of infertility. The many factors that may cause infertility complicate diagnosis and prevention for you and your vet alike.

Unobserved Heats

In Ben's herd, one glaring problem was "silent heats." After I examined his cows, it was apparent that the problem was really *unobserved heats*. The signs of unobserved heats are obvious (even though this statement appears to be contradictory). *You will see no heats*. In these cases, the cows usually have long average days open and calving intervals, with a normal number of services. The average days open usually is more than 150 days. Production is lower than you think acceptable.

The causes for unobserved heats range from cows not cycling to dairymen not observing cycling cows. Unfortunately, most dairymen blame cows for not coming into heat when the fact is that most unobserved heats are because not enough time is devoted to heat detection.

A review of DHIA data on calving intervals showed that dairymen are losing twice as many days from not detecting heats as from actual conception failures.

According to an Ohio study, the average days open is 154. About 41 of these days are from missed heats and 13 are caused by conception failures. Clearly, heat detection is a bigger problem.

Of course, some cows don't exhibit heats even with adequate observation. Many environmental conditions contribute to this. Icy lots or smooth concrete floors make cows fearful of falling during mounting or standing. Stress from high humidity or extreme heat also can make cows lethargic. Overcrowding can also be a problem because of social interaction and difficulty in observing individual cows. Overcrowded herds are also often stretched thin on labor because of overtaxed facilities and resources.

Tie-stall barns also hinder good heat detection because cows can't move freely. Tie stalls are a particular problem when cows aren't turned out regularly for exercise. Unobserved heat may even be a problem in herds observed closely for signs of heat. In these herds, the focus of the vet should be on whether the cows are cycling. These herds may be on a poor plane of nutrition. Herds in financial distress are often shortchanged on feed, particularly in late winter. The efficiency of heat detection usually drops in these situations.

A new cowside method of evaluating heat detection is available for use in the parlor. The milk progesterone test can be used to test milk weekly for changes in progesterone levels. If the levels rise, the cow has cycled. A steep drop in progesterone usually indicates that the cow is in heat; she should be bred the third consecutive day that the level is low.

Failure to detect heat is seldom related to one facet of your management, and usually stems from a number of deficiencies. This was also the case with Ben. He relied on his hired man to watch for heats while he fed the cows. The hired man, Orville, had no incentive to watch for heats, other than Ben's growling at him. Orville needed a financial incentive to spot cows.

A specific time should be set aside to observe cows for heat. Most cows mount others only once every 15-20 minutes, so the observation period must be at least 20 minutes long.

Many of Ben's cows weren't cleaning properly at freshening and had uterine infections. Because of the uterine disease, many didn't cycle. Also, when Ben's cows did cycle, they didn't show good heats. Ben's attitude was, "What's the use in watching because they don't do anything?"

I found that his barnlot was slippery from years of scraping. The herd's diet also had a mineral imbalance. Once these were corrected, the cows were very active sexually.

Correcting Unobserved Heats: Treatment for unobserved heats is determined by the underlying causes of the problem. As in Ben's herd, the situation in your herd must be considered from a number of angles. To improve heat detection, PGF_{2a} can be used to synchronize heats so you can anticipate heat. When many cows are already showing sexual activity in a herd, other potential breeders are much more likely to show strong heats. If a cycling cow is the only open cow and the others are pregnant, she is likely to show only a very mild heat.

Heat detection devices are a great aid. In large herds, gomer bulls and androgenized cows are good teaser animals for heat detection. Androgenized cows are made by giving them large doses of the male hormone testosterone.

Sometimes progesterone implants are used in cows not showing heat. These are removed after 10 days, and the cow then shows heat. The implants are good but are too expensive for routine use in commercial cows. These implants should only be used under the direct supervision of your vet.

Nutritional Deficiencies

There is a broad category of cows in which heat detection and the diet are good but the cows aren't cycling. Improving the levels of all dietary minerals often helps herds with problems in heat detection, assuming cattle are conscientiously observed for signs of heat. In my opinion, there aren't many herds that won't benefit from improved mineral nutrition.

Low dietary phosphorus not only adversely affects heat detection but also impairs absorption of other minerals that also influence heats. Low dietary magnesium can cause cows to not cycle. Beta-carotene, vitamin A and cobalt also are necessary for normal cycles. Vitamin E and selenium are important in the heat cycle in that deficient cows tend to have no heats or are cystic.

Heifers often do not cycle immediately after calving. They usually have a negative energy balance because they are still growing, milking and trying to reproduce. They can't meet all necessary functions. In many cases, they are also crowded out at the bunk by mature cows.

Heat can usually be detected by palpating cows not showing signs of heat. If the cow is cycling, a corpus luteum (CL) is often palpable. If these cows have a CL, the methods of heat detection should be investigated. If no CL is palpated on weekly visits, then the cow's nutritional status should be checked. Slippery lots and other environmental factors also should be checked.

Metritis

Many repeat breeders become problems because of unresolved uterine infections (metritis). Most cases of metritis are usually easily spotted. The cow has a white vulvar discharge. The tail is raised and sometimes her back is arched (Fig 1). Many cows have a foul smell. Most infected cows don't cycle. If they do cycle, the AI gun will be flecked with pus when the gun is withdrawn at breeding.

Figure 1. A cow standing with her tail raised and back arched usually has irritation in her pelvis. This often occurs from calving or rough calving assistance by an inexperienced person. Sometimes it is seen also with urinary infections or hardware disease.

In acute cases, the cow has a fever and may be off feed. Some cows become dehydrated. In these cases, metritis may be life threatening.

In either of these situations, the cow becomes a repeat breeder. On a herd basis, the records reflect increased days open, longer calving interval, and increased numbers of services to conception.

The causes of metritis are primarily related to a cow's ability to clear or purge her uterus after calving. When a cow has an effective defense mechanism cleaning up her uterus, bacterial infections don't persist after calving. Some cows have an overwhelmed uterine defense mechanism from anatomic defects that cause constant contamination with manure and urine. Cows with poor nutrition are very susceptible to metritis.

Pyometra is an extraordinary infection that cows may get after calving. The uterus fills up with pus and mucus, and the cow doesn't cycle. It may occur while the cow is in the luteal phase of her cycle. As long as the cow doesn't cycle, the infection persists.

Metritis is usually diagnosed on rectal examination by an experienced palpator. Many dairymen can diagnose metritis by watching closely for pus discharging from the vulva. A word of caution: vulvar discharge of pus may signify a normal clearing of the uterus after freshening. If the vet is in doubt or suspects infection, a sample can be taken for culture of the infection. Occasionally, a biopsy or sample of tissue may also be used for the diagnosis.

It is best to prevent persistent metritis. Nearly every cow has some degree of uterine damage after calving. If she has been fed correctly and is in good health, metritis usually is cleared by the uterine defense mechanism. *Uterine infusion should not be a routine part of getting cows ready to breed*. Antibiotic infusions in the uterus are a poor choice as compared to improving management factors to help prevent metritis.

Few dairymen put enough effort into monitoring their herd's ration. If the dietary mineral level is optimal and the cow calves in a reasonable environment, she should not need infusion. This doesn't mean, however, that you should throw your infusion rods away. There are instances where antibiotic infusion is indicated. It may be used after therapy to make the cow cycle. The drug of choice for infusion is usually 1-2 million units of penicillin diluted to a larger volume. Be careful to observe your vet's recommendation on milk withholding after these infusions.

Ovarian Disease

Ovarian disease is also a common reason for reproductive failure. A common problem of high-producing cows is small inactive ovaries.

These cows don't have any heats and may be thin. On rectal palpation, the ovaries are small and have no palpable structures. The cause is a lack of energy. The cow may not be getting enough energy in her feed, or she may be giving more milk than feed consumption can support. Nutritional deficiencies that interfere with energy utilization can also cause ovarian problems. They include deficiencies of phosphorus, magnesium and cobalt. Illnesses that reduce feed intake also cause small, inactive ovaries. Commonly, cows with displaced abomasums take several months to get bred after surgery. Debilitating illness, such as acute mastitis or footrot, can also cause this syndrome. Treatment revolves around preventing the energy deficit. Increasing the nutritional plane hastens recovery.

Nearly every dairyman wants to know what to do with cows that have cystic ovaries or delayed ovulation. Ben was no exception. He wanted the magic shot that would make these cows breed 60 days after calving.

Two types of cysts are commonly seen in cows. With the follicular cyst, the cow doesn't ovulate and has a large, thin-walled vesicle on the ovary. The luteal cyst may cause a partial ovulation and luteinization of the follicle on the ovary.

Follicular cysts are more likely to cause continuous heats or heats every few days. Cows with luteal cysts are likely to not show heat. Cystic cows sometimes suck air in the vagina, leading to a condition called pneumovagina or "windsucker." Cows that have been cystic a long time often get a high tail head and may exhibit heat when another cow is in heat. Some may even act like a bull. Luteal cysts may cause cows to have longer cycles, such as 23-24 days. These are diagnosed by observation of frequent heats or upon rectal exam detecting a mature follicle immediately after ovulation should have occurred.

The causes of cystic ovaries may be a lack of luteinizing hormone (LH) and nutritional deficiencies. A lack of vitamin E and selenium, especially in the dry period, predisposes mature cows to cystic ovaries. Some cows that are beta-carotene deficient may also have more cystic ovaries. Moldy feeds can greatly increase cystic ovaries and other problems. Older cows are predisposed to cystic ovaries because of persistent uterine disease. Heredity also plays a role in cystic ovaries. But we cannot blame cystic ovaries on heredity without first investigating other possible causes.

Injections of estradiol (ECP) in oil in the period just after calving increases cystic ovary problems in high producers. Cows coming in heat soon after calving also have a higher incidence of cystic ovaries. Cows in heavy milk production that are superovulated for embryo transfer are also prone to cystic ovaries.

Though Ben was only concerned with getting the magic shot, he had several management problems. He had moldy feed containing an estrogenic mycotoxin in the silo because his silo doors leaked. His cows were also vitamin E-selenium deficient. They were not only more prone to cystic ovaries, but frequently had retained placentas.

Some ovarian cysts can be diagnosed with the milk progesterone test. A milk assay may reveal that the cow didn't produce high progesterone levels 6 days after heat. Low progesterone indicates she formed no luteal tissue, so she is likely cystic. You can also diagnose cystic ovaries by the history of an irregular heat cycle. In other instances, she has a high tail head and is a windsucker.

There are a number of treatments for cystic ovaries and delayed ovulation. Routine treatment can be accomplished with gonadotropin-releasing hormone (GnRH). Some vets also manually rupture the cyst if it can be done with little damage. *Gently* rupturing a follicular cyst apparently causes no problems for the ovary. Other compounds, such as human chorionic gonadotropin (HCG), pituitary-like hormone (PLH), and other LH-like compounds are also effective. The problem with these compounds is that cows eventually become immune to them. PGF_{2a} should be used for luteal cysts to dissolve the luteal tissue and cause a heat. Progesterone therapy works in cows with repeated episodes of cystic ovaries. Progesterone implants during the luteal phase are a real aid to the embryo donor cow that is prone to cysts. Implants can prevent cysts on valuable cows that continue to be cystic after repeated treatment. If delayed ovulation is the problem, GnRH can be given in the very early part of the heat; the cow is then bred on schedule.

Cows may not conceive because of the after effects of the previous calving. One of these is ovarian adhesions. These develop from prolonged metritis. A serious infection ascends the uterus and Fallopian tubes, and ultimately spreads to the ovaries. The result is scarring (adhesions) of the Fallopian tubes and ovaries.

Rectal exams in which the palpator tries to remove a CL often result in an adhesion. If only one ovary is involved, the cow will continue to cycle but likely will only conceive when the ovulation occurs on the unaffected ovary.

Ovarian adhesions are diagnosed on rectal exam. If the adhesion involves both ovaries, permanent infertility is likely to result. There is no reliable therapy for this condition. The cow will be fertile only if the adhesions are few or one ovary is free. If adhesions are on both ovaries, most of these cows are sent to slaughter.

Another condition involving the ovaries is a persistent corpus luteum (CL). Some vets believe that a CL palpable on an ovary while a cow is in heat represents a persistent CL. If this condition is suspected, a milk progesterone assay will quickly resolve the debate. Low progesterone levels indicate there is no functional CL. If a functional CL is found by measuring progesterone while a cow appears to be in heat, it can easily be resolved by waiting 7 days after the heat and giving the cow PGF_{2a}.

Vaginal Disease

Vaginal conditions can be involved with infertility. Cows that suck air into the vagina (windsucking, pneumovagina) often have structural defects between the vulva and anus. Some of these cows with pneumovagina may also be nymphomaniacs with cystic ovaries (Fig 2).

Pneumovagina has several causes. Cows that have had a difficult calving may be torn from the vulva to the anus. This most commonly occurs in heifers. Older cows that have calved several times often have stretching of the vulvar muscles. When these muscles sag, they let air and manure into the vagina. In both cases, the vagina becomes inflamed and the cow tends to strain. The result is that the windsucking gets worse. Cystic ovaries also predispose cows to pneumovagina.

You can diagnose pneumovagina by hearing the windsucking. Experienced palpators can diagnose it during palpation or at breeding. A cow with pneumovagina has a large balloon-like structure where the vagina and cervix are normally located. When they encounter this, many of my clients are puzzled and call me to breed the cow. This balloon-like structure is actually the vagina full of air.

Usually the treatment involves repairing the defect causing the condition. The procedure can be very minor, but can be very involved if calving trauma was severe. Cows that have pneumovagina because of cystic ovaries recover when the cystic condition is corrected.

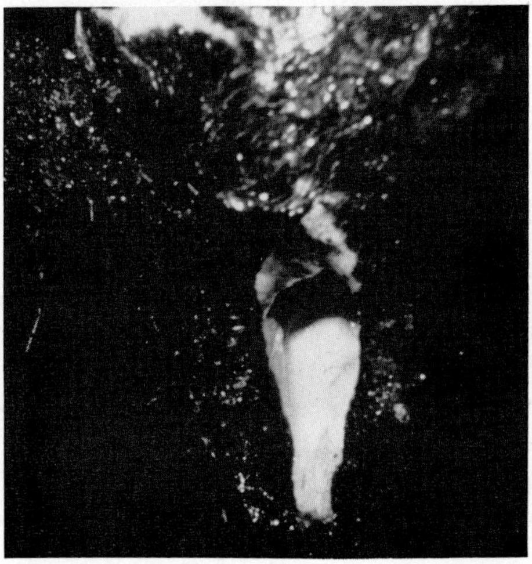

Figure 2. Air and dirt may enter the vagina and cervix because of poor confirmation or injury around the pin bones and vulva at calving. These cows are prone to vaginal infections and chronic infertility. Your vet can repair these defects with a surgical procedure called Caslick's operation.

Inflammation of the vagina (vaginitis) often can reduce fertility. Vaginitis may cause signs similar to those of other infections, such as metritis. There is a vulvar discharge of pus. In acute cases, the tail is raised and the back arched. Urination may be painful. The membranes lining the vulva are red and inflamed. Little blisters may be interspersed with the redness.

Every cow has vaginitis after calving. The vagina has been stretched and bruised from calving. It takes several days to weeks for the cow to heal. Various infections also cause vaginitis.

Some cows in Ben's herd also had vaginitis. It appeared to be low grade and probably was only a small factor in his general reproduction problems.

Vaginitis is diagnosed by looking at the vagina by opening the vulva or viewing it through a sterile tube (speculum) with a flashlight. Identifying the type of infection requires special cultures. Transportation is a problem if the lab is in a distant city.

Vaginitis is treated with douches of tetracyclines or mild antiseptics. Natural service should be avoided because these infections are a type of venereal disease. This means that the bull will become infected and spread the infection to other cows. When using AI, the infection is self limiting. One additional precaution should be taken. A double sheath is put over the breeding rod so the infection is not carried into the uterus, which would cause the cow to be temporarily infertile. If the double rod is used, the infection is not transported to the uterus and becomes self limiting in the vagina; the cow will conceive normally.

When the infection has been identified as *Hemophilus* vaginitis, many dairymen vaccinate for *Hemophilus*. While vaccination may be warranted for other problems, *Hemophilus* vaccination for vaginitis doesn't appear to help.

Other Causes

There are many unexplained circumstances in which cows are infertile. However, you shouldn't categorize a breeding problem as unexplainable without a thorough investigation of the factors already mentioned.

High-producing cows naturally have more breeding problems. Ben always lamented that his problems never happened to the bad cows, just his good ones. This is often the case. The best cow is under the most stress. She is likely not getting enough nutrients to meet her needs. She simply can't eat enough. Her mineral requirements are higher than the rest of the herd's because of her high production. She needs special attention because she is working hardest.

This cow is milking at a high level and losing weight. Upon palpation, her ovaries are usually smooth and hard. She may not be cycling. If she is cycling, her uterine tone is poor because nutrient levels are meeting milk production requirements but not reproductive requirements.

Treatment of this cow requires "tincture of time." Her nutrient levels must be corrected to eliminate the energy deficit. Eventually, she will breed if every effort is made to correct the deficits.

Dietary deficiencies often cause cows to be repeat breeders. Initially the cows become infertile because of the deficiency. Once these cows get into this rut, correcting the deficiency doesn't quickly solve the problem. Their uterine defense mechanism isn't easily changed. Cows that freshened with the deficiency don't breed satisfactorily. Meanwhile, newly freshened cows on the corrected diet breed efficiently. Consequently, it is much easier to correct a deficiency before the cows manifest the signs of deficiency.

Copper deficiency causes infertility in cows that freshened with the deficiency. A lack of copper causes a poor immune response. Part of this immune response occurs in the reproductive tract. The uterus heals poorly and conception is unsatisfactory. In addition, copper is related to vitamin A. If copper is lacking, vitamin A levels are low, and cows may not show good heats.

Cows can be anemic with cobalt deficiency and are energy deficient because of the inability to fully use what they are fed (see Chapter 19).

Selenium deficiency is a major reason for infertility. Cows deficient in selenium and vitamin E tend to be cystic and not show good heats. This is one of the most important micronutrient factors in the Midwest for getting cows bred.

Phosphorus is important for trace mineral absorption. Phosphorus shortages cause cows to not come in heat. This is a very important nutrient to monitor in all dairy production areas of the world.

Herds with poor breeding efficiency must be monitored for stress. Excessive environmental heat and humidity cause great problems with conception rates. Excessive cold and icy conditions stress cows and increase breeding problems.

Overcrowding stresses the herd, and heat detection becomes a problem. Many cows may not show good heats because of the "pecking order." Environmental stress can be detected by correlating the breeding records to the weather and housing situation.

This situation is corrected by eliminating the underlying causes. Reduce the stress of the "pecking order" by providing more room.

Provide shade and shelter. Forced ventilation may be required in confinement housing. Cows under hot, humid conditions may need extra dietary potassium. Feed density may need to be increased by cutting back on high-moisture feeds. Feeding during the cool of the night also is beneficial. Improve fly control so that cows aren't constantly pestered by flies.

Sometimes a specific cause of infertility cannot be identified. The cows may have normal cycles and normal heats periods. Other cows may have abnormally long cycles or long heats. These cases must be treated according to their signs.

Often these cows have been fresh a long time. Most are overly fat and have had problems calving. They may have chronic infections. Usually they have a poor uterine defense mechanism and scarring in the uterus. Uterine disease may be associated with delayed ovulation and cystic ovaries.

Regular palpation through the cycle by your vet may indicate the diagnosis and correct treatment. If this doesn't dictate the appropriate treatment, these cows should have a single-egg (not superovulated) embryo transfer flush. When the cow has been bred, the flush is done 6-9 days later. At that time, the diagnosis and prognosis can be made. If no egg is recovered, the Fallopian tubes may be blocked. If an egg is recovered, it can be determined if fertilization occurred and if embryonic development is normal. If a normal embryo is recovered, it can be put in a recipient heifer. The superior genetics of this cow can then be continued through new generations. If a cow's genetics do not warrant an embryo flush, she's usually culled.

12

Detecting Pregnancy

Various physiologic changes occur when a cow conceives. The ovulated egg from the ovary is fertilized by a sperm 14-24 hours *after* heat. The fertilized egg travels slowly down the Fallopian tube to the uterine horn and begins to divide: first into 2 cells, then 4 cells, then 8 cells and 32 cells. By about the fifth day it consists of 64 cells.

The embryo arrives in the uterus and continues to divide, relying on secretions from the uterine glands for nutrients. By 42-45 days after conception, the embryo has implanted itself in the uterine wall, with a direct attachment to the lining (endometrium) for its nutrients.

Every normal ovulation forms a corpus luteum (CL) at the site of ovulation on the ovary. The CL gradually starts producing progesterone. By the sixth day after ovulation, the uterus is under the optimum influences of progesterone from the CL. The CL is maintained as long as the cow remains pregnant. If, by 17 days after ovulation, there is no pregnancy, the CL dissolves and a new heat cycle is started.

Pregnancy causes changes in the ovaries, as well. It's common for pregnant cows to have follicular growth and continue to develop follicles that mature, then shrivel up and disappear. Some of these cows may have enough follicular growth so they continue to show heat periods despite pregnancy.

Since your goal is to see that every cow has a calf every year, you must accurately detect pregnancy as soon as possible after conception. *Failure of a cow to return to heat is not a reliable sign of pregnancy.*

Early detection of pregnancy also gives you an opportunity to detect problems in cows that didn't conceive. And you'll have the jump on treating infertility problems. In general, prompt pregnancy detection improves milk production because it reduces the average days open. Each cow has a shorter calving interval, and more chances for lactations in her lifetime.

How to Diagnose Pregnancy

Pregnancy can be diagnosed in a number of ways.

Ballottement

With ballottement, you stand on one side of the cow and apply your fist against the abdomen (Fig 1). Gently push your fist in and out against her flank, where the abdomen extends outward the farthest. In advanced pregnancy the calf suspended in the uterus swings back and forth, and bounces against your fist.

Ballottement has advantages and disadvantages. There's no cost in "bumping" for a calf. Any dairyman can do it quickly, without any preparation. The disadvantage is that you can't detect early pregnancies. In fact, this method is reliable only after the sixth or seventh month. Even then, it is not totally reliable. Finally, neither you nor the vet can detect problems related to the pregnancy.

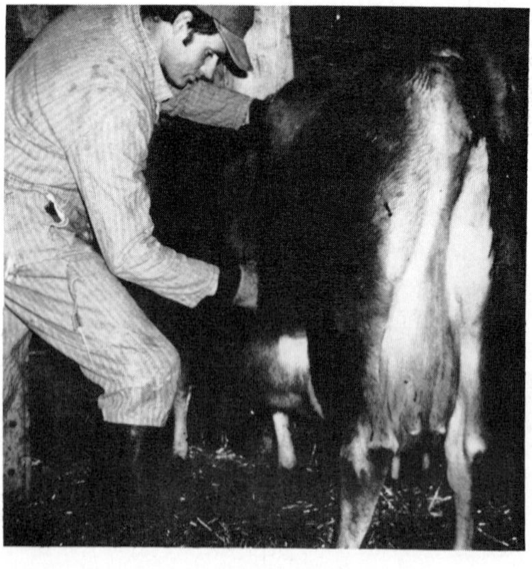

Figure 1. An ancient method of diagnosing pregnancy is by ballottement, or "bumping," for a calf. In advanced pregnancy, the calf can be felt when you move your fist deep into a cow's abdomen.

Milk Progesterone Assay

The milk progesterone assay is marketed in a kit designed for use by you or your vet (Fig 2). It measures a cow's milk progesterone level, which is directly proportional to the amount of progesterone in the blood. A cow has a high level progesterone in her milk during the luteal portion of her cycle and when she's pregnant. You use the test 19 days after the last breeding. If the milk progesterone level is low, the cow is open. The test reliably shows that a cow is open but cannot differentiate between a cow in the luteal phase and one that's pregnant.

The advantages of the assay test kit are numerous. It's easy to perform and readily available to dairymen. It also tells you whether a cow's open just 19-24 days after breeding. There's no risk to the cow, and you can complete the test in about 15 minutes.

An obvious disadvantage is that the assay doesn't diagnose pregnancy reliably, as mentioned earlier. It's also expensive if you're only testing a small number of cows. You must exactly follow the test procedures, and you can't take shortcuts. Many dairymen aren't willing to take time away from all of their other duties to run regular assays. If a dairyman would spend the money used for a pregnancy diagnosis kit on heat detection, I wonder if this might work as well.

"Whup! I let go of her tail, didn't I?"

Rectal Palpation

Rectal palpation is about 98% accurate in diagnosing pregnancy. It is almost always done by your vet and requires considerable training

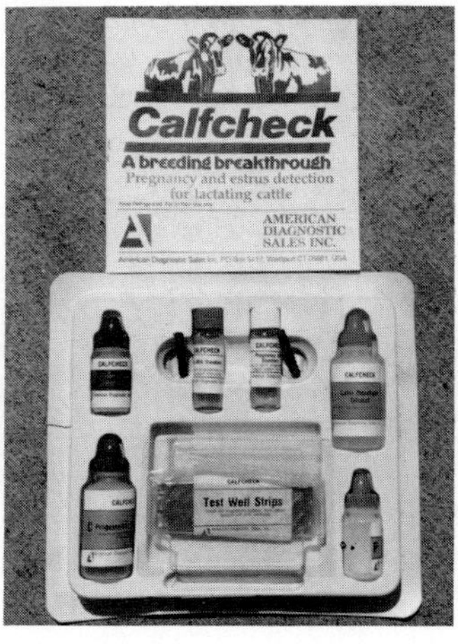

Figure 2. The milk progesterone test can sometimes determine which cows are open 19 days after the last breeding. Though the test is useful in diagnosing pregnancy, false readings often occur because the high progesterone level during pregnancy is identical to a cow's high progesterone level in mid-cycle.

and experience. The untrained palpator may injure the cow, causing rectal lacerations and possible resorption of the fetus.

The vet evaluates a number of structures inside the cow in this process to confidently diagnose pregnancy. In early pregnancies, he notes a tense, thin-walled, fluid-filled uterine horn. Skilled vets can confirm a pregnancy at 30-35 days. The palpator can also feel the amniotic vesicle as he gently runs his hand over the uterine horn. Using the "fetal membrane slip," the vet uses his fingers to gently pull on the uterine horn, and then let it slip back. This is an especially touchy procedure, and an inexperienced palpator can easily cause a cow to resorb the fetus. Experienced palpators do the slip in the body of the uterus, rather than in the horn. I prefer to use other means of diagnosing pregnancy.

Pregnancy can also be diagnosed by feeling the fetus itself, about 70-75 days after conception. Pregnancy can also be confirmed by feeling the attachments of the cow's uterine lining to the membranes of the calf. These cotyledons confirm pregnancy after 65 days.

Between days 90 and 125, "fremitus" is another tip-off to a confirmed pregnancy. The cow's middle uterine artery greatly expands to accommodate the increased blood flow to the unborn calf. A distinct "whirring" or vibration can be felt on rectal palpation, much like you'd feel if you grasped a garden hose and felt the water rushing through it.

By 5-6 months of pregnancy, the fetus and other structures have moved forward over the pelvis in deep-bodied cows, so the only structure you can palpate is this middle uterine artery. It's an excellent way to diagnose pregnancy in late pregnancy.

There are a number of advantages of rectal palpation. As mentioned earlier, an experienced palpator can accurately diagnose pregnancy 98%

"Forget your glasses again, huh, Doc?"

of the time. A skilled veterinarian can also use rectal palpation to diagnose other problems. If he finds that the cow isn't pregnant, he may be able to determine what's wrong. The vet can also evaluate pregnant cows that are showing heat. About 10% or more of cows in heat are pregnant. Other cows that you think are pregnant and showing heat may really have other reproductive problems.

Early diagnosis lets you treat the cow early, rather than wasting time waiting for her to freshen. Take advantage of the opportunity to pick your vet's brain while he's doing palpations. In my experience, this has always been a good time to talk about managing the herd.

There are some disadvantages of rectal palpation. Depending on the palpator's expertise, you may lose pregnancies through embryonic death and fetal resorption. Another disadvantage is that you probably can't do this procedure yourself, except in the last stages of a pregnancy. So you'll need to call on the vet for help, and sometimes you have to wait longer than you like for him or her.

13

Maintaining Pregnancy

Before I tell you about the conditions that threaten pregnancy, I'll try to help you visualize what's going on inside a pregnant cow, from fertilization until freshening.

The Cow's Reproductive Tract

The uterus is divided into 2 horns similar in shape to a ram's horns. At the tip of each horn, connected by the Fallopian tube, is an ovary. The cow has a right and left ovary connected to a right and left uterine horn. The ovary ovulates the egg, the egg is fertilized at breeding, and pregnancy occurs on the same side as the ovulation. It's impossible for one horn to maintain a pregnancy if the embryo is in the horn on the opposite side of the ovulating ovary. This is a critical point in embryo transfer technique. For example, you'd need to place the embryo in the recipient heifer's right horn if she ovulated on the right side.

The uterine horns are made up of several layers of tissue. The innermost layer, called the endometrium, contains the glands and blood supply to nourish the fetus to term. Underlying the endometrium is the myometrium, which is the muscular layer of the uterus. These muscles contract to transport sperm to the ovary and move the calf during delivery. Enclosing the endometrium and myometrium is the serosa, which is the protective layer exposed to the abdominal environment.

For the first several weeks that the embryo is in the uterus, it receives its nutrients from the uterine glands. The fetal membranes begin to develop and gradually the uterine caruncles form. These caruncles are button-like projections that supply nutrients to the unborn calf. By 65 days, the cow supplies all of the nutrients to the fetus via the blood supply to these caruncles. The nutrients diffuse across the caruncles into the calf's receptors, called cotyledons. The caruncle and cotyledon match like an electrical plug and an electrical outlet. A cow may have as many as 75-120 caruncles.

The Developing Fetus

Various things can go wrong as the embryo develops into a fetus and then into a calf. By day 22 after fertilization, a crude heart has formed, and it's beating. By day 25, the spinal cord is functional. Defects that occur during this period are usually life threatening, so the embryo may die, be expelled at the next heat, or be absorbed without any external signs in the cow.

The developing fetus is protected principally by a membranous sac containing amniotic fluid. The fluid acts as a shock absorber to external blows. If the fetus develops a defect while this membrane and fluid are forming, excessive fluid may be produced, up to 8-10 times the normal volume. This condition is called hydrops amnion or hydrops allantois, depending on the origin of the fluid.

Gradually the placenta develops. Nutrients in the dam's blood pass through the placenta to the fetus, and fetal waste material passes back into the dam's blood. As mentioned above, the placenta is where the caruncles match up with the cotyledons to allow this exchange of nutrients and waste.

Fetal Abnormalities and Death

Freemartinism

One developmental problem at this stage of pregnancy is called freemartinism. In cases of twins, where one twin is a heifer and the other is a bull, the female twin is sterile about 90% of the time. She's called a freemartin, and her reproductive tract is abnormal (Fig 1).

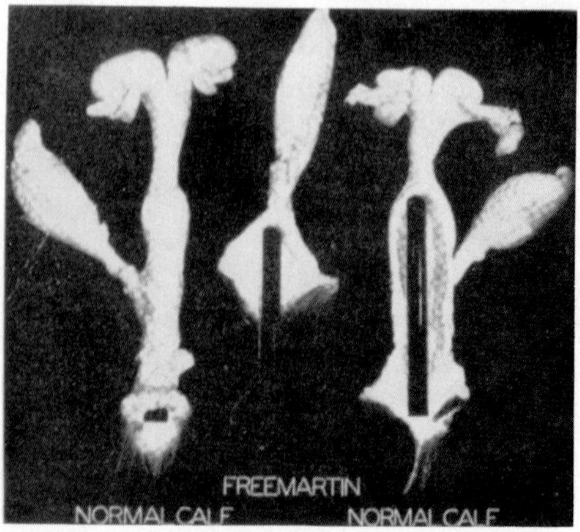

Figure 1. The vagina of freemartins typically is underdeveloped. A veterinarian can diagnose freemartinism by inserting a 6-inch test tube into the vagina. The tube usually can pass nearly its entire length into the vagina of a normal heifer calf. In a freemartin, the tube can pass only about half its length.

Your vet can often diagnose a freemartin by looking at a suspect heifer's vagina right after birth. If he has doubts about the normality of her reproductive tract, raise her and reevaluate her by rectal palpation at 12-15 months of age.

Embryonic Death and Resorption

Embryonic death and fetal resorption are terms you're probably familiar with. These events usually become evident when a cow returns to heat 3-6 weeks after being diagnosed pregnant. Your first reaction might be that the vet called her wrong. If your vet is a skilled palpator, this is seldom the case. Instead, some defect in the developing fetus or its dam caused termination of the pregnancy.

Fetal resorption may occur for a number of reasons. First, the fetus may be defective in some way, so it does not develop to term. Either the embryo fails to develop properly after fertilization, or it is resorbed because the uterine environment is not appropriate to maintain a pregnancy.

Second, mold toxins (mycotoxins) in your feed may prompt fetal resorption. Examples of these mycotoxins are zearalenone and vomitoxin.

Deficiencies of vitamin E and selenium have also been reported to cause embryonic death and fetal resorption. These deficiencies probably are among the more common reasons for interrupted pregnancies.

Older cows that have had a number of calves or have had serious uterine infections frequently have a lot of uterine scarring or fibrosis. These cows may not have adequate uterine glands to nourish the developing embryo. So the embryo dies and is resorbed.

Vibriosis (*Campylobacter* infection) is another cause of fetal death. This is common in herds that use a bull. Other causes of fetal death include heat stress, bovine virus diarrhea (BVD) infection during the first 3 months of pregnancy, any systemic disease, or hormonal imbalances. This is especially true with older cows. In these cows, the CL fails, progesterone levels drop, and the embryo is resorbed.

Your vet can diagnose embryonic death and fetal resorption either by repeated rectal palpation or milk progesterone assay. The best way to avoid this condition is to correct the problem that caused this loss. For example, you may need to provide an older cow with supplemental progesterone so she can maintain her next pregnancy.

Fetal Mummification

A fetus can die and become mummified when the blood supply in the uterus is cut off. Either the cotyledons and caruncles are separated, or a disturbance of blood flow interferes with diffusion of nutrients between these 2 structures. The fetus dies but does not decay because it is in a sterile environment. The cow doesn't abort, and the cervix doesn't open, so there's no chance for bacteria to cause infection. Gradually, the cow absorbs the fluids from the fetus, causing it to shrink, harden and eventually resemble an Egyptian mummy.

You may suspect this condition when a pregnant cow doesn't show signs of advancing pregnancy or fails to "spring" near her due date. The cow won't calve on time if she has a mummified fetus. Occasionally she may eventually come back in heat and expel the mummy (Fig 2).

A skilled palpator can diagnose a mummified fetus during rectal palpation, but sometimes it takes 2-3 palpations over a month or more to confirm the diagnosis.

Prostaglandin therapy can be given so the cow will expel the calf. Or you may choose to send the cow to slaughter. Some dairymen decide to have a C-section done on a really valuable cow. Unfortunately, the chances of full recovery for breeding aren't good, so cesarean section should be viewed as a measure of last resort.

Fetal Maceration

A macerated fetus develops when the fetus has died and bacteria have entered the uterus through an open cervix and caused the fetus to decay. Often parts of the fetus are expelled from the womb. Some parts, especially bones, may remain and become embedded in the endo-

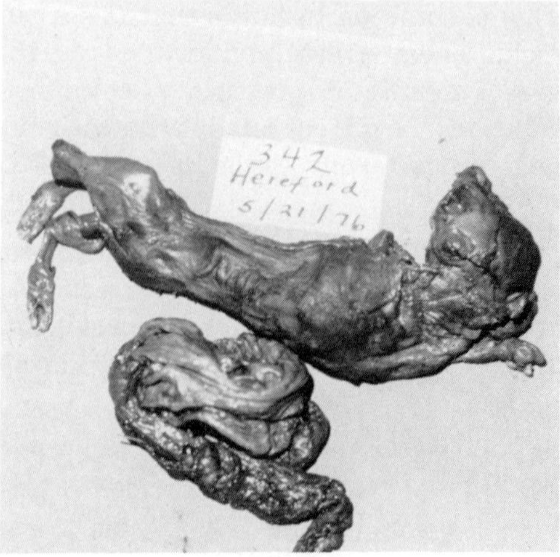

Figure 2. If a fetus dies in the uterus and the cervix remains closed, the fetus loses its fluids and becomes mummified. The cow may be diagnosed as pregnant and never freshen. (Courtesy of Dr. T.L. Blanchard, Texas A&M University)

metrium. In these cases the potential for serious infection is great. If the bones are trapped in the endometrium, they act like an IUD and cause permanent damage and sterility (Fig 3).

This condition may be associated with trichomoniasis or torsion of the uterus just in front of the cervix, which prevents fetal expulsion. It is diagnosed by rectal palpation. Pyometra is often evident, too. In some instances, this is life-threatening to the cow. Surgery can be done to remove the infectious debris, or the cow can be sent to slaughter.

Abortion

Abortion can occur anytime during pregnancy. It is often associated with fetal death (Fig 4). Any systemic disease that stresses the cow may

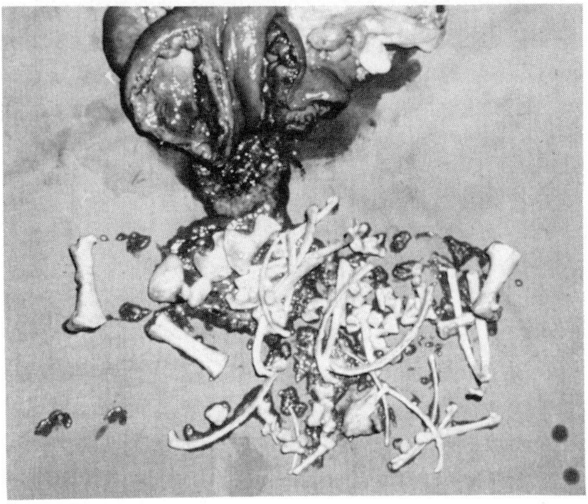

Figure 3. If the fetus dies, the cervix dilates and the fetus putrefies. Occasionally the cow survives and the bones and hide of the calf remain embedded in the uterine wall. This type of fetus is called a macerated fetus. (Courtesy of Dr. T.L. Blanchard, Texas A&M University)

Figure 4. An infection or developmental abnormality during pregnancy usually results in abortion. This calf was aborted at about 7 months of gestation. (Courtesy of Dr. R.G. Elmore, Texas A&M University)

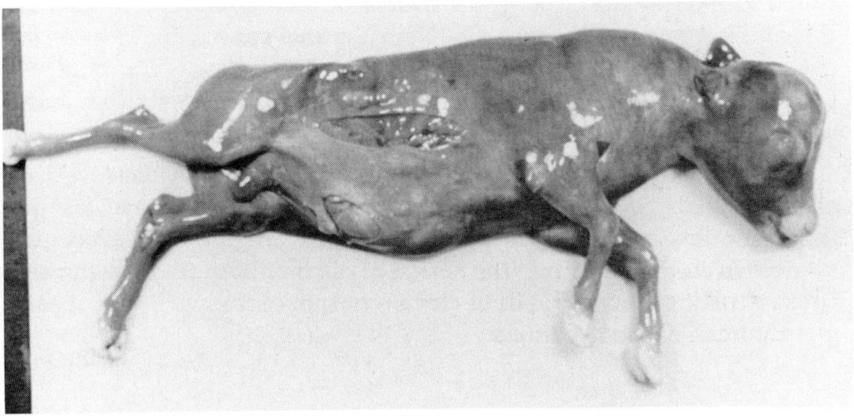

result in abortion. The most common causes of abortion are mycotoxins, infectious bovine rhinotracheitis (IBR) and bacterial infections. Leptospirosis, bovine virus diarrhea (BVD) and *Hemophilus* infection are additional causes. Brucellosis, chlamydiosis and listeriosis are occasionally the cause. Cows also may abort because of twins or a defective calf. One twin may impinge on the other fetus's circulation. The second twin dies or produces cortisone from the stress. This initiates labor and the cow delivers twins before they are prepared to survive. Defective calves also can initiate labor.

Mycoplasmosis may cause abortion more often than we suspect. This is because the organisms are difficult to culture and identify. Vibriosis and trichomoniasis also can cause a loss. Numerous poisons are known to terminate pregnancy. They can act by several methods and can be life-threatening to the dam or the unborn calf. Another possible action is to compromise the blood supply to the caruncles, stressing the unborn calf. This stimulates the unborn calf or its dam to secrete cortisone, which triggers uterine contractions and subsequent abortion. Nitrate poisoning is noted for causing abortion by interfering with the unborn calf's oxygen supply, which creates the abortion-causing stress.

If you administer cortisone to a springer cow for udder edema or other conditions, you may also trigger the calving mechanism, and the calf is delivered from the uterus early.

In addition to a number of systemic diseases, localized infections can cause termination of pregnancy. Placentitis, which is a localized inflammation of the placenta, can cause abortion. This commonly occurs in older cows that have had chronic uterine infections and scarring. Placentitis interferes with proper exchange of nutrients between the maternal caruncles and the fetal cotyledons. Fungal infections also act on the caruncles and cotyledons, and are likely to terminate a pregnancy between the third and seventh months.

If one of your cows overeats on grain concentrate, she gets a condition called rumen acidosis or grain overload. The condition is also life-threatening and could cause abortion. Various genetic defects can also cause abortion, but they are not common. A genetically faulty fetus is usually aborted by the sixth to the eighth month of gestation. Rarely will a cow abort from a fall or severe blow.

Hormonal imbalances, such as low maternal progesterone levels, can also cause abortions. An older cow is more likely to have low progesterone levels, which fail to maintain the pregnancy. Certain substances in feed that mimic the action of hormones in the cow may also cause abortions. Some types of clover contain excessive levels of estrogen that can cause abortions.

Abortion can lead to a number of complications for the cow. Most cows that abort have retained placentas. Because the cow wasn't prepared for calving, the placenta is usually tightly adhered to the uterine caruncles. This usually results in a serious infection and fibrosis or scarring of the endometrium. If the fibrosis is localized, the cow can sometimes compensate in the next pregnancy with the rest of her endometrium. But many cows have difficulty becoming pregnant after aborting.

14

Diseases Affecting Reproduction and Production

In a number of ways, disease affects a cow's ability to reproduce. It may have a localized effect on the fetus or on the cow's uterus. A disease may have a secondary effect on the uterus from systemic involvement; stress may stimulate a rise in the cow's internal cortisol levels. It also can have an effect on the blood supply to the fetus.

Disease also impairs milk production in a number of different ways. Systemic diseases affect the cow's homeostasis or equilibrium. They also affect a cow's appetite. Disease stress raises cortisol levels, which can suppress milk production.

The following pages contain an easily read summary of facts on dairy cattle diseases. At the end of the chapter is a checklist of the basics in preventing disease.

Anaplasmosis

Background: Found primarily in Wisconsin. Infectious but noncontagious. Caused by red blood cell parasite. Spread by biting insects (mosquitoes, horse flies) or by contaminated syringes and needles.

Signs: Related to destruction of red blood cells. Acute anemia. Marked reduction in performance. Early fever, pale color to gums. Later, muscular weakness, depression, dehydration and off feed. Respiratory distress. May not be able to absorb adequate oxygen because of low red blood cell count. Potential for abortion.

Diagnosis: Observe clinical signs, then blood smear exam under a microscope, blood tests, autopsy.

Treatment: Long-acting tetracycline for nonlactating animals (also used as preventive during insect season). Recommend vaccination every 3 years but can cause problems with an antibody factor in colostrum, prompting complications in newborn calf (comparable to RH factor in human babies).

Prevention: Isolate carriers by blood test and start program to eliminate carriers. Insect control.

Bovine Respiratory Syncytial Virus (BRSV) Infection

Background: New disease. May be in bovine respiratory disease complex, causing pneumonia and shipping fever. Apparently widespread. Caused by virus. Must be differentiated from IBR, BVD, pasteurellosis. Spread by nasal secretions, coughing, sneezing.

Signs: High fever (106-108 F). Coughing. Runny nose. Eye discharges. Rapid breathing. Appear alert in spite of high fever.

Diagnosis: Based on paired blood tests taken 2 weeks apart. Virus culture difficult. Possibly by autopsy.

Treatment: Supportive care. Antibiotic therapy and relieve respiratory distress.

Prevention: BRSV vaccinations.

Bovine Virus Diarrhea (BVD)

Background: BVD antibodies usually found in nearly every blood serum sample. Caused by virus. Several strains. Some severe infections (called mucosal BVD) caused by 2 strains at same time. Controversial disease for veterinarians to give specific recommendations.

Signs: Rough haircoat, dehydration, reduced milk production or cow dries up. Maybe fever, depression, poor appetite, weight loss.

In *pneumonia form*, ulcers on gums and muzzle, respiratory infection with nasal discharge, harsh dry cough. May not have diarrhea. Can cause death.

Diarrhea form seen more often than other forms. Loose stool to watery diarrhea, mucus strings, blood clots and undigested feed in stool. May have scanty gray, pasty stool, ulcers on lips, gums or side of tongue and esophagus, nasal discharge, watery eyes. May die. In severe case (mucosal disease) calf may have pneumonia form of BVD at same time.

Neurologic form causes brain damage (impaired balance) in unborn calves exposed to BVD virus between third and sixth month of pregnancy. Born normally, appear healthy except unable to stand and nurse. May suck while lying down but won't recover. Other affected calves born blind or with cataracts.

Laminitis form causes inflammation of sensitive quick in foot. Cow has stilted gait. May get coronary band cracks. Reluctant to walk.

Pinkeye form may not show systemic signs of BVD but calf has runny nose, watery eyes, conjunctivitis. Eventually cornea ulcerates and calf is blind.

Reproductive form causes abortion in first trimester. May cause infertility in cows harboring BVD virus in uterus.

Immunologic form means that a developing calf is unable to produce immunity to infection if exposed between days 90 and 165 of pregnancy. Can carry virus with no response to disease. Exposure to another BVD virus type causes mucosal disease (a very severe form) and death.

Diagnosis: Based on clinical signs plus isolation of virus or paired serum testing on blood samples. Change in titer indicates BVD exposure. Postmortem exam.

Treatment: No specific cure, so give supportive antibiotics to control secondary infections, intravenous fluids, electrolytes, intestinal astringents, aspirin or flunixin.

Prevention: Start vaccinations with killed vaccine when calf 4-8 months old. Follow with booster in 2 weeks; repeat 1 month before breeding. Then annual vaccinations. Killed vaccine safest but offers poorest protection. Live vaccine protects best but can spread BVD to unborn calves of the herdmates.

Brucellosis (Bang's Disease)

Background: High incidence in southern states, totally eradicated in many others. Can infect people (undulant fever). Three strains of organism found in different food animals. Enters through eyes, mouth, nostrils. In pregnant cow, organism invades caruncle-cotyledon junction and interferes with transfer of nutrients to calf. Spread to feed and water by aborted fetuses, placenta and fluids from infected cow. Infected cattle can be carriers.

Signs: Bacterial infection in blood (bacteremia). Bacteria also shed in milk. Drop in milk production. Abortions or stillborn calf after fifth month of pregnancy. Sometimes calf born alive but weak. Fresh cow has retained placenta. Uterine infection. Often becomes repeat breeder and long-term infertility case.

Diagnosis: Blood tests (card, plate, tube). Also milk ring test (MRT) or brucellosis ring test (BRT) performed by health departments on milk. Culture.

Treatment: No cure. Some cows eventually develop immunity and carry calf to term.

Prevention: Vaccination about 60-70% effective. Calfhood vaccinations with live-modified vaccine (reduced dose strain 19) between 4 and 8 months in some states, between 3 and 6 months or 2 and 6 months in other states. Strain 19 reduced dose available for adult cows in a few states. Where Bang's is still a problem, eliminate natural service. Vaccination must be done by your vet.

Chlamydiosis

Background: Diagnosed in problem herds where vague disease observed. Caused by organism found in birds and some mammals. Spread by eye and nasal secretions. Nutritional deficiencies greatly increase susceptibility. Common reproductive disease in people caused by different strain from the disease in cattle.

Signs: Abortion in 5-10% of pregnant animals. Pneumonic form seen in calves. Arthritis in 3- to 6-mo-old calves (swollen knee joints). Pinkeye, diarrhea. Bulls in stud can develop infection in sex glands.

Diagnosis: Based on clinical signs and paired blood serum drawn 2 weeks apart to look for changing titers. Culture (difficult procedure).

Treatment: Controversial. Tetracyclines and spectinomycin work well on pneumonic form. Correct copper or other nutritional deficiencies, overcrowding, other stresses.

Prevention: Vaccine gives immediate protection but lasts only 6 weeks so must repeat. Vaccine readily available is cat vaccine, used extensively on cattle. Other vaccines available from traveling drug salesmen not well studied, may be waste of money.

Clostridial Infections

Background: Group of diseases caused by closely related family of organisms. Deadly!

Botulism organism can grow in high-moisture ryelage or round bales. Can kill.

Tetanus organism causes lockjaw. Carried in intestinal tract of horses. Causes infection when invades cut or puncture wound of animals or people. Produces powerful toxin that causes gradual paralysis of all skeletal muscles. Those around jaw often affected early.

Blackleg organism causes muscle inflammation. Similar to gangrene in people. Common in calves on pasture, in feedlots. Evident after animal bruised. Early lameness, depression, anorexia, rumen inactivity, fever 104 F. Animal reluctant to move, swelling with air under skin in muscle groups (shoulder, rump or thigh). Later, tremors, difficult breathing, coma, and death in 12 hours or less.

Malignant Edema and Gangrene: Start from a wound infection. Shock and sudden death. Affect all animals without regard to age, sex or breed but usually more common in healthy animals at least 3 months old. Fresh cow may develop this infection in uterus, show no clinical signs and die quickly.

Black Disease and Red Water Disease: Infection in liver more common in cattle that eat snail-infested forage grown in poorly drained soils. Snail is intermediate host for liver flukes. Spores of bacteria survive well in contaminated soil. Probably spread by feces or urine of infected animals, also transported by water. Infection starts with liver insult from parasite migration, abscesses, chemicals, injuries, plant and mold toxins. Signs are incoordination, staggering, depression, poor appetite, rapid breathing, reluctance to move. Later, appear heavily tranquilized, breathe heavily, occasionally grunt. Rapid death with acute infection within 18-36 hours (red water disease) or 60-100 hours (black disease). High fever (104-106 F) and wine-colored urine with red water disease. Normal temperature with black disease.

Entertoxemia Types A, B, C and D: Types C & D most common. Usually affect calves. Cause enterotoxemia, an acute intestinal infection often called "overeating disease." Good-looking animal may eat ravenously and later be found dead. In less acute cases, animal drops, convulses, throws head back and dies in a couple of hours. May have diarrhea.

Diagnosis: Dependent on signs of toxemia, sudden death. Lab tests. Postmortem exam.

Treatment: Difficult. Few animals saved. Large doses penicillin at site of gangrene for blackleg and red water disease. For enterotoxemia, evacuate intestinal tract and give activated charcoal and sodium bicarbonate. Intravenous antitoxin also.

Prevention: 7-way or 8-way vaccine that includes tetanus. Give to young stock, heifers, dry cows. Start with 2 doses 2 weeks apart, then 1 dose annually to all animals 3 months or older. Sometimes need type C & D overeating disease vaccine for younger calves.

Hardware Disease

Background: Caused by animal eating metal or foreign object in feed. More common in cow that's an indiscriminate eater. May puncture reticulum (chamber of stomach), penetrate lungs and cause pneumonia; penetrate diaphragm and liver and cause liver disease; or penetrate heart sac and cause heart disease. Can also penetrate muscle and skin beneath cow's chest and cause no permanent ill effects. May trap object and wall off with scar tissue, then cow gets sick at next calving because of abdominal straining with labor.

Signs: Pain in elbows and withers. Grunt when walking or breathing. Pulse in jugular vein. In acute cases, cow is off feed, usually slight fever, drop in milk production.

Diagnosis: Listen to rumen activity. Observe clinical signs. Blood count. X-ray or ultrasound.

Treatment: Give magnet to keep metal from penetrating further (Fig 1). Give rumen stimulants to get intestinal tract going again. Retrieve hardware via stomach tube. Minimize secondary infections with supportive antibiotic therapy. Surgery to retrieve hardware. May have to slaughter.

Prevention: Put magnets on feed-handling equipment. Give each cow a magnet. Be very particular about the feed sources.

Hemophilus Infection

Background: Venereal disease transmitted during natural service breeding. Can also invade mucous membranes of nostril and mouth on air and in secretions. More likely in stressed cattle and those with nutritional deficiencies. Several forms:

Pneumonic form: Fever (up to 106 F). Calves off feed, hacking cough, runny nose, eye discharge.

Nervous form: Off feed, nervous behavior, incoordination, fever. May soon die with meningitis.

Reproductive form: Vaginitis, metritis, infertility, occasional abortion. Unborn calves may be stillborn.

Arthritic form: Sore, swollen joints. May cause drop in milk production without any other signs.

Figure 1. Because cows eat everything in their feed, they must have a magnet in their stomachs to pick up any metal they eat.

"I think that magnet you put in her stomach might have been just a little too strong."

Diagnosis: Difficult. Culture. Organism is fragile, killed by freezing, sample transport. Draw blood samples 2 weeks apart to compare titers though may not be definitive. Postmortem exam.

Treatment: Antibiotics (especially penicillin) and fluids. Supportive care except with vaginitis, then douches helpful.

Prevention: Semi-annual vaccination helps prevent pneumonic, arthritic and meningitis forms but not against reproductive form.

Infectious Bovine Rhinotracheitis (IBR)

Background: Common viral disease. Probably found on nearly every farm. Spread in air or by nasal discharges, or contaminated gloves, nose leads, halters. May take 60 days after exposure before clinical signs seen. Animal may carry virus in brain cells for life, shedding whenever stressed. Cows commonly infertile for several months after outbreak. Several forms:

Pneumonic form: Fever (105-106 F), loss of appetite, depression. Later, coughing, sharp decrease in milk production, snotty nose, ulcers in mouth and nasal cavities, nose scab ("red nose"), labored or rapid breathing.

Pinkeye form: Increased tearing, pus discharge from eye, conjunctivitis, cloudy ulcer on cornea. Looks like pinkeye.

Reproductive form: Called infectious pustular vulvovaginitis (IPV). Common in heifers. Pain upon defecation or urination, white ulcers and pus on vaginal floor.

Abortive form: Fetus expelled immediately after IBR exposure, or virus may incubate and calf aborted months later. Usually cow aborts 3-5 weeks after infection.

Encephalitic form: Nervous signs, quick death.

Intestinal form: Diarrhea, especially in young calves. Hard to distinguish from BVD.

Diagnosis: Based on clinical signs. Isolation of virus from fetus, placenta, or nostrils of affected cow. Blood tests to check titers.

Treatment: Supportive care, antibiotic therapy to prevent secondary infection, fluid therapy.

Prevention: Maintain closed herd (see Chapter 4). Start vaccinations at 6 months. Repeat 1 month before breeding, then annually. Intranasal vaccine effective. May be given to calves at birth or soon after. Modified-live intramuscular vaccine unsafe for pregnant cattle. Modified-live only safe as intramuscular injection. Killed intramuscular vaccine safe in pregnant and open cattle, does not shed, rarely causes allergic reactions and lethargy. Provides less immunity than modified-live products. One chemically altered injectable modified-live product does not cause abortion.

Johne's Disease

Background: Also called paratuberculosis. Found in dairy areas of the United States, more common in Wisconsin and New York. Techniques for diagnosis still somewhat undeveloped. Caused by bacteria. Calves pick up infection in first 30-60 days of life (and up to 6 months) without signs, carry disease for years, then break with diarrhea at 2-6 years. Spread through contaminated feces. Nutritional deficiencies (cobalt, selenium) increase susceptibility.

Signs: Chronic, debilitating, incurable diarrhea. Cows waste away over time, die of malnutrition. High incidence of chronic mastitis.

Diagnosis: Difficult. Manure culture takes up to 19 weeks. New blood test (AGID) correlates well with culture only if animal has diarrhea. ELISA test (not refined yet) may soon show carriers not yet showing signs and are culture negative — will be valuable as cattle prepurchase test. Biopsy of cow's rectal wall (false negatives occur). Examination of lymph node taken from intestinal tract (culture and biopsy). Intradermal test similar to TB test, but useless. Intravenous Johnin test. Watch for temperature rise and change in blood count after cow injected with Johnin protein (false negatives occur). Postmortem exam.

Treatment: None effective or justified. Only suppress diarrhea. Cow is still shedding. Slaughter affected animals.

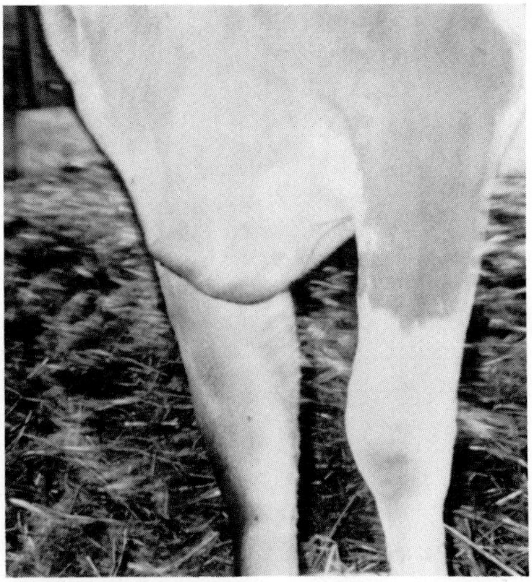

Figure 2. This Guernsey heifer was vaccinated for Johne's disease. Note the increased thickness of her brisket. Such heifers often are positive to TB tests because they have been sensitized to tuberculin by the Johne's vaccine. Consequently, they are unlikely to pass interstate health requirements for shipping without a special permit from the state veterinarian.

Prevention: Clean up manure. Eliminate calf exposure to manure by pulling from affected dam immediately after birth. Pasteurize colostrum. Scrape away 3-4 inches topsoil in dry lots where cows calve — Johne's organism can live in soil for years. Raise trace mineral levels (especially cobalt) to optimum levels. Improve ration to improve immune status. Vaccinate calves 1 month after birth. (Prevents new cases but can't eradicate organism from affected herd). Vaccine injection can cause permanent lump in brisket (may be as large as a softball). May make calf test positive on TB test (Fig 2). Can also culture manure on all cows in herd and cull. Repeat every 6 months. Still may miss cows incubating disease and not yet shedding organisms.

Lameness

Background: Feet and leg problems common in dairy cattle, especially in herds intensively managed on dry lots and concrete.

Foot Rot: Bacterial infection rots away tissue between toes. Common in spring when ground is muddy. Sharp stones and irritating substances can work their way between toes, or get a start in hot dry weather when hard hooves crack (Fig 3).

Laminitis (Founder): Inflammation causing growth changes in foot. Can be caused by excessive energy in ration without adequate fiber or buffering, and by systemic disease, such as BVD or nitrate poisoning. Common in cattle housed on concrete. Also follows puncture wounds or lacerations of sole, or systemic infection, such as metritis.

Heel Cracks and Sole Separations: Usually result of nutritional deficiencies (copper, zinc) (Fig 4).

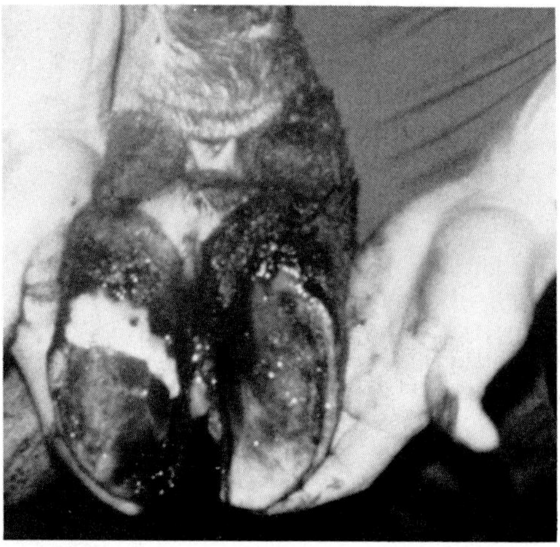

Figure 3. This cow has foot rot or a puncture wound. A tract extended into the bony part of the foot and erupted above the coronary band. (Courtesy of Dr. Glenn Hoffsis, Ohio State University)

Faulty Conformation: Usually passed through genetic influences from sire and dam, but conformation can deteriorate if animal spends life standing on concrete.

Signs: Lameness, lethargy, failure to compete for feed, lower production, may refuse to get up. Sore or crack between toes or on bottom of foot. May have soft soles or heel cracks. May see draining tracts from bottom of foot, sole abscess, or cracks extending up into sensitive part of foot. Splayed foot (weakened ligaments and bone). Crooked toe from dislocation. Deformed toes or deformed leg bone.

Diagnosis: Observe gait; examine feet. Sometimes x-ray necessary for valuable cow. Can reveal hidden nail or stone (Fig 5).

Treatment: If herd problem, evaluate nutrition and environment. Make management changes. Antibiotic therapy for infected lesions. Cut

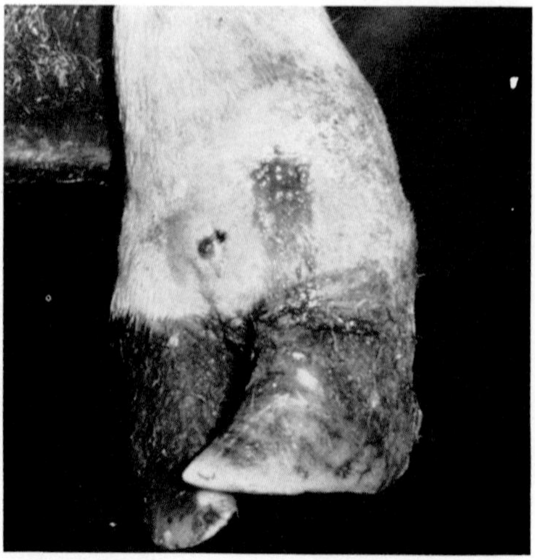

Figure 4. Erosions of the heel, heel cracks, and an extremely soft foot are typical in cows with a copper deficiency. Copper is necessary for sound hooves.

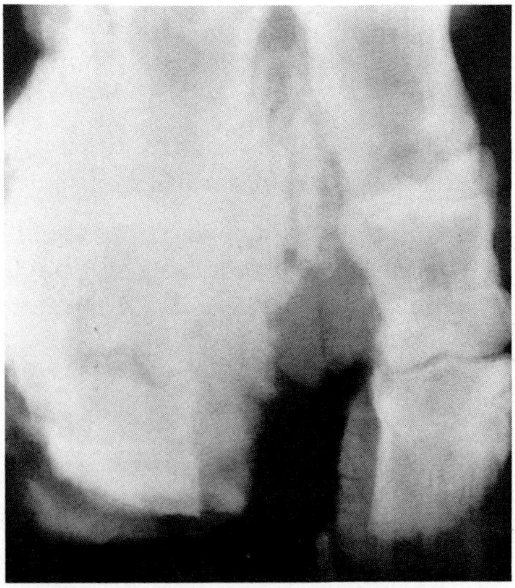

Figure 5. On this radiograph, the left toe has normal bone. The right toe has a moth-eaten appearance. This indicates extensive inflammation in the right toe. (Courtesy of Dr. Glenn Hoffsis, Ohio State University)

away dead, infected tissue. May have to glue wood block to good toe to relieve pressure on sore toe. Flush affected toe. Chemical cautery (iodine crystals and turpentine or heat) to clear up draining tracts. Chronic case, can amputate one toe if not on slats.

Prevention: Regular trimming. Balanced ration. Remove potential sources of foot injury.

Leptospirosis

Background: Warm weather disease. Increasing incidence in some areas. Caused by several types of bacteria called spirochetes. Five types commonly seen in dairy cattle (7 isolated in United States, 100 worldwide). Organism feeds on dead, decaying tissue. Found in rivers, ponds, stagnant water, sewage. Spread by urine when cattle drink from contaminated water sources. Organism causes liver disease and damages kidneys. Cattle can continue to shed organism for months after infection. Also found in rodents, spread to cattle when rat urine contaminates feedstuffs. Spread also by direct contact of urine to cattle skin and mucous membranes and during natural service with urine or semen. Can invade abrasions, cuts in skin or invade uterus via blood stream. People are susceptible to disease and develop persistent flu-like symptoms.

Signs: Mild to severe. High fever (105-106 F), poor appetite. Milk appears yellow and clotty. Dairyman treats for mastitis with no response. Then ropy milk disappears, milk output depressed. Calf can become persistently infected if organism invades dam's uterus late in pregnancy — calf will likely die if not aborted.

Diagnosis: Paired blood tests 2 weeks apart to determine if infection is active. Isolate organism from infected animal. Postmortem often fruitless.

Treatment: Dihydrostreptomycin for cow that persistently sheds organisms in urine but doesn't show clinical signs.

Prevention: Killed vaccines available for 5 types. Annual vaccinations in closed herds every 6 months if herds have history of lepto. Keep cattle away from stagnant water supplies, streams, ponds, sewage. Control rodents.

Teat Lesions

Background: Caused by pox virus, vaccinia virus, pseudocowpox, warts, frostbite, injuries, lacerations.

Signs: Prolonged milking time due to poor letdown. May be kicky. Mastitis. Skin may peel off. Frostbitten teats turn black, ends swell shut and teat may even slough off. Blisters may indicate systemic virus moving through herd. If nursing calves, they may have blisters on muzzle (Figs 6-9).

Diagnosis: Clinical signs. May need to isolate virus if a herd problem.

Treatment: Often isn't rewarding. Keep teats clean, soft (topical ointments). Chlorhexidine teat dip and ointment to fight virus. Milk cows infected with pox virus last. Disinfect teat cups between cows. Wear rubber gloves, rinse hands and utensils in disinfectant washwater between cows.

Prevention: Reduce chances of cows' getting teat abrasions, chapping. Eliminate muddy areas where cows drag udders. No good vaccines for warts. Sometimes success with a vaccine vet can make from warts taken from teats. Dry teats and use no teat dip in extremely cold

Figure 6. This heifer has a herpesvirus infection called mammillitis. Note the severe inflammation of the skin of the teats. Most of the skin may slough off. (Courtesy of Dr. Ralph Farnsworth, University of Minnesota)

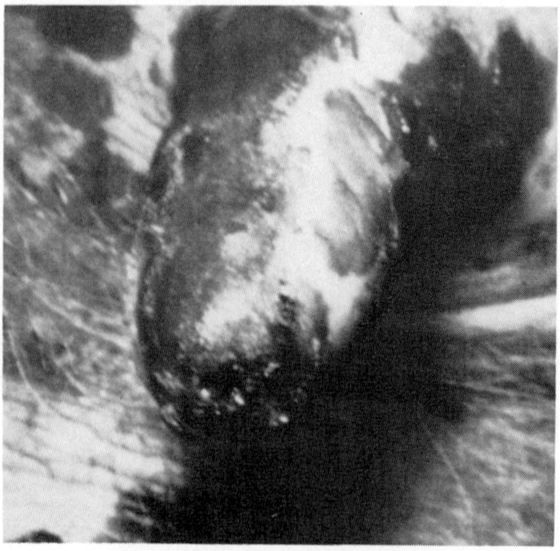

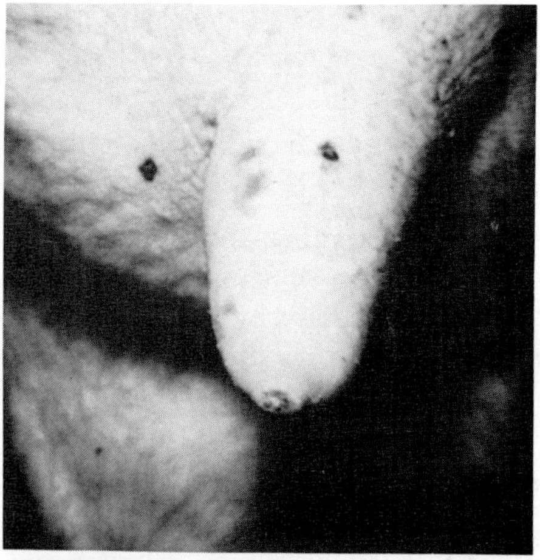

Figure 7. This cow has a typical infection from pseudocowpox virus. Note the pinpoint lesions scattered about the teats. (Courtesy of Dr. Ralph Farnsworth, University of Minnesota)

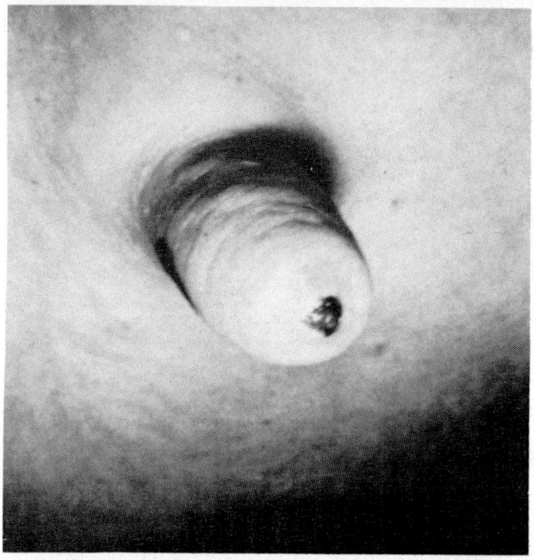

Figure 8. A cow with teat end erosion is likely to get mastitis because of the damaged teat sphincter. This type of lesion may be caused by high vacuum, overmilking or trauma. (Courtesy of Dr. Ralph Farnsworth, University of Minnesota)

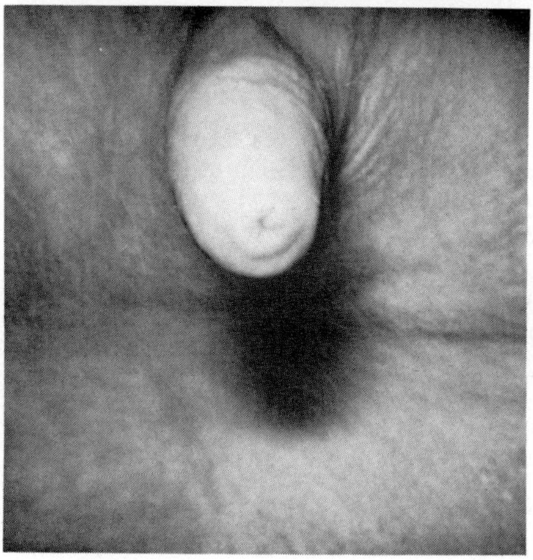

Figure 9. Note the smooth ring callus on the end of this teat. In years past, lesions like these were thought to be caused by faulty milking machines or overmilking. Now we know that a callus like this is not caused by any infection or irritation and is of no consequence. (Courtesy of Dr. Ralph Farnsworth, University of Minnesota)

weather. Keep cows (especially fresh cows) out of weather with wind chill of 20 F or lower.

Leukosis

Background: Nearly half of US dairy cattle exposed. Caused by virus carried by biting flies and contaminated injection needles. Often spread by transmission of blood from one cow to another. Can also spread slowly through herd by nasal secretions and in air. Calf can be exposed before birth, or pick up leukosis antibody in colostrum.

Signs: Subtle. Most cases go unnoticed. Blood test necessary before export to many countries. Positive result eliminates animal as candidate. Cattle exposed to leukosis show positive blood test. Second form of virus causes tumors around abomasum (displaced or ulcerated abomasum), in chest cavity (jugular pulse, congestive heart failure) or spinal cord (posterior weakness, paralysis). Thymic form in 6- to 20-mo-old calves. Causes tumors in brisket, swollen lymph glands in shoulder area. Skin form rare, looks like ringworm.

Diagnosis: AGID blood test. Cattle may test positive without showing signs. Observation, palpation of tumors. Blood count (high number of one type of white blood cell). Postmortem exam.

Treatment: None.

Prevention: Eliminate chances for blood contamination between animals. Use clean needle on each animal, then discard. When dehorning calves, dip dehorners in disinfectant between calves. Control biting insects. Quarantine or eliminate any infected animals. Isolate newborns from infected dams. Pasteurize colostrum from leukosis-positive dams. Avoid natural service. Test all new herd additions.

Listeriosis

Background: Caused by organism common in nature. Found in soil and manure, ensilage during winter season. Readily passed in milk products, so a public health hazard unless milk is pasteurized. Three forms:

Nervous Form: Causes encephalitis or brain infection. Localizes in one side of brain. Cow circles, can only walk in one direction ("circling disease").

Abortive Form: Exposure during pregnancy increases risk of abortion. Cow exposed late in pregnancy may have weak calf. Serious health hazard to pregnant women and young children.

Septicemic Form: Newborn and young calves have high fever, become seriously ill, often die.

Diagnosis: Clinical signs. Postmortem exam and culture.

Treatment: Systemic antibiotics. Young stock may benefit from low-level tetracyclines in feed.

Prevention: Killed vaccines have variable success. Live vaccines are a human health hazard. Best to prevent soil and manure contamination of feed, especially ensilage. All dairy products for human consumption must be pasteurized.

Lumpy Jaw and Wooden Tongue

Background: Two distinct clinical diseases, both caused by organisms in oral secretions of cattle. Infection occurs when mouth is cut or abraded with foxtail awns or splinters in forages.

Signs: Lumpy jaw causes enlargement of jawbone, bony tissues. Wooden tongue causes acute onset, enlargement of soft tissue. May also cause abscesses, respiratory difficulty, draining tracts (Fig 10).

Diagnosis: Palpation of lesions, x-rays. Microscopic evaluation of discharge.

Treatment: Systemic organic iodine and tetracyclines. Some vets use penicillin-streptomycin. Surgery to remove affected tissue is rarely used.

Prevention: Cull affected animals to prevent further contamination of environment.

Mycoplasmosis

Background: Organisms difficult to isolate. Can cause mastitis, arthritis, infertility, abortion, vaginitis, metritis, pneumonia.

Figure 10. Lumpy jaw is a bacterial infection of the bony tissue of the jaw. It usually develops from bacteria invading a puncture wound or ulcer in the mouth. A cow with lumpy jaw should be culled from the herd to avoid contaminating the feed bunks with the organism. (Courtesy of Dr. Elaine Hunt, North Carolina State University)

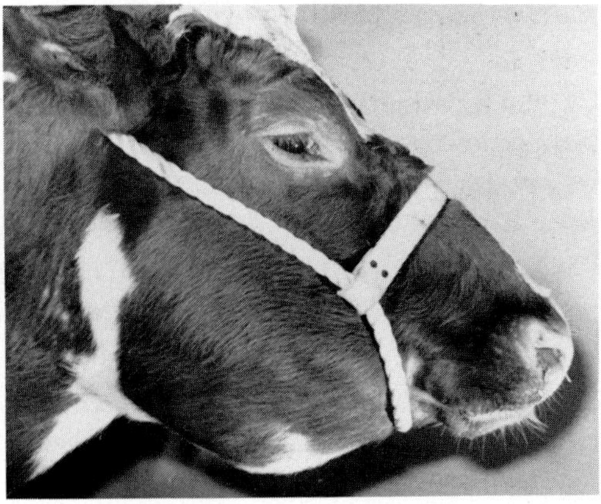

Diagnosis: Difficult. Cultures necessary, require special handling by lab personnel. In certain cases, blood tests are also useful.

Treatment: Tetracyclines, tylosin, lincomycin-spectinomycin.

Prevention: Improve nutrition. Maintain closed herd. In a few cases, making a vaccine from mycoplasma isolated from herd has helped. Avoid natural service. With vaginitis form, use double rods on breeding sheaths (AI).

Pasteurellosis

Background: Organism is normal inhabitant of upper respiratory tract. Infections caused by stress (poor ventilation, low immunity, shipping, previous viral infections). Transmitted by nasal secretions, coughing.

Signs: Coughing, rapid breathing, nasal and eye discharges, fever, pneumonia. Frequently affects cattle at same time as other respiratory diseases.

Diagnosis: Culture from lungs of calf that died from pneumonia and from aborted fetus (abortion rare). Characteristic lesions. Postmortem exam.

Treatment: Antibiotic therapy.

Prevention: Live vaccine effective but doesn't give complete immunity. Killed vaccines may make calves susceptible to pneumonia. Improve ventilation and calf housing (see Chapter 4).

Salmonellosis

Background: Caused by bacteria that are eaten or picked up from other infected animals. Organism found in wet, cold facilities with stressed animals. Also in effluent from waste spread on fields and manure contamination of feed.

Signs: Severe diarrhea, then dehydration, depression, off feed. Calves frequently get pneumonia, arthritis. Some strains cause abortion. Once established in herd, milk output drops, newly fresh cows are especially vulnerable. With peracute form, cows die suddenly without exhibiting signs.

Diagnosis: Tentative diagnosis by observing clinical signs. Blood count (white blood cell count drops). Manure culture. Postmortem exam.

Treatment: Rapid intravenous fluids. Antibiotics (controversial because they often cause carriers afterward).

Prevention: Vaccine made from organisms from your cows. Keep cow and calf facilities clean. Eliminate manure contamination of feed. Maintain closed herd.

Trichomoniasis

Background: Usually found in herds using natural breeding. More common in beef than dairy cattle. Caused by protozoa found in reproductive tract of cattle.

Signs: Pus discharge from vulva. Pyometra, irregular heats, repeat breeders. Abortions common in first trimester. Rarely more than one sign at once. Low-grade infection undermines reproductive success of herd over time. Often only cows affected at once.

Diagnosis: Culture reproductive tract, aborted fetus, bull's sheath for organism.

Treatment: None for dairy cows. Cows that abort develop immunity. Some new drugs stop infection in valuable bulls.

Prevention: Use AI. Buy semen from studs approved by National Association of Animal Breeders. A new USDA vaccine is also available.

Udder Sores

Background: Caused by bacteria or fungi on damp udder surfaces. Common between leg and udder. Excessive swelling in first-calf heifers. May slough skin between leg and udder from pressure sores. Pressure sores in first-calf heifers should not be confused with fungal or bacterial infection of these areas. Udder canker and sores can also be caused by subcutaneous worm the cow acquired by lying in contaminated dirt lots.

Signs: Chronic sore slow to heal. Foul smelling, spreads if not treated. Can cause loss in production.

Diagnosis: Clinical signs.

Treatment: Clean sore with surgical iodine scrub. Topical ointment to dry up. If sore between udder and leg, may need chemical cautery. Treat subcutaneous worm by giving cow chemical agent that kills larval form migrating in udder tissues.

Vibriosis (*Campylobacter* Infection)

Background: Caused by bacteria. Seen principally where natural service used. Virgin heifers highly susceptible. Can spread through untreated AI semen. Normally contracted from infected bulls and contaminated bedding.

Signs: Irregular heats (30-35 days long). Virgin heifers become repeat breeders and have vulvar discharge. Abortions possible between fourth and seventh months.

Diagnosis: Culture of vaginal mucus, fluids from aborted fetuses. Identify suspect bull by breeding to virgin heifer.

Treatment: Uterine infusions rarely successful. AI is best because the heifers gradually develop immunity.

Prevention: Vaccinate heifers before breeding with bull and again at freshening. Use antibiotic-treated semen from reputable AI stud. Exposed animals develop some immunity.

Winter Dysentery

Background: Occurs in winter months in northern US, Canada. Herd may not break with it if exposed earlier. Not clear how it is transmitted; some indication it is spread by birds. Rarely fatal. Caused by a coronavirus.

Signs: Diarrhea, drop in production, fever. Diarrhea may be blood-tinged. Disease can be explosive if herd never before exposed. Affects young milking heifers most severely.

Diagnosis: Based on clinical signs, elimination of other possible causes of diarrhea (salmonellosis, BVD). Based also on herd history. Characterized by explosive onset and short duration (1 week).

Treatment: Intestinal astringents, supportive treatment. Seldom requires antibiotics.

Prevention: No vaccines. Maintain closed herd. Monitor visitors. Insist salesmen scrub boots before and after visit. Wear clean coveralls (see Chapter 4).

A Checklist for Preventing Disease

- Maintain a closed herd.
- Avoid travel to farms affected by disease.
- After visiting another farm, change clothes and boots when you return home.
- Try not to lend out manure spreader or other equipment that is used in your dairy lot.
- Consult with your vet and evaluate the risk of adding new animals to your herd.
- Isolate newly acquired cattle and observe in quarantine.
- Examine health history of new animals before allowing them to enter your herd.

- Schedule vet exam of new individual, start vaccination program.
- Provide good rations and clean water. With advice from vet and/or nutritionist, feed to reduce vulnerability of your cows.
- Reduce sources of stress: provide shade and cool water in summer, protect cows (especially fresh cows) from extreme temperatures, eliminate overcrowding.
- Check design of barn lots, eliminate sources for injuries.
- Practice good sanitation in cow lot.
- Protect feed supply from contamination.
- Review proper milking techniques. Use teat dips routinely. Wear rubber gloves.
- Milk infected cows separately, preferably last.
- Sanitize milking equipment thoroughly.
- Develop and follow vaccination programs for cows, heifers, calves.
- Deworm heifers regularly.
- Use AI. Buy semen from a reputable stud.
- Use double rods when breeding cows with a Cassou gun.
- Be alert to all abnormalities in your herd. Check with your vet for disease problems.
- Isolate sick animals, milk separately and discard milk if necessary. Wash and disinfect hands after giving treatment.
- Provide foot bath in isolation area to disinfect boots. Use only treatments recommended by your vet.
- Use separate instruments and utensils on sick animals to avoid spread of disease.
- Disinfect equipment and isolation area after sick animals recover, leave vacant for several days.
- Promptly burn, arrange for pickup or bury dead animals and aborted fetuses, retained placentas, fluids and bedding, especially if disease is contagious.

15

Preventing and Controlling Mastitis

Mastitis is the most costly disease of dairy cattle. The National Mastitis Council estimates that current losses from mastitis amount to $200 per cow annually. There are 10 million dairy cows in the United States, which means that you and your fellow dairy farmers are losing $2 billion a year from mastitis.

Losses occur in several ways. The losses in milk production amount to 70% of the costs of mastitis. Milk discarded after treatment with antibiotics amounts to 8% of the total. Drugs and veterinary expenses cost another 8%, and another 14% is from death and premature culling. By far the largest loss from mastitis is decreased milk production in cows with subclinical (mild) mastitis.

Dairymen must also be concerned about public health and milk marketing. When cows get mastitis, the somatic cell count goes up in the milk tank. This creates a bad image in the public eye. The taste of milk also deteriorates as somatic cell counts rise. Increased levels of lipase in the milk contribute to rancidity. High somatic cell counts in milk affect milk composition. There are decreases in lactose, casein and fat. There are also increases in sodium, chloride and whey protein, which cause an increase in milk pH. All of these changes contribute to a decreased shelf life of milk in the store.

In addition, there is a public health concern about antibiotic residues in milk. Cows with mastitis frequently require treatment with antibiotics. Historically our industry has a recurring problem with antibiotic residues (Fig 1). Whether we like it or not, the public wants a consistently pure, wholesome product in the dairy case.

As you know, mastitis is an infection of the mammary gland. Mastitis has been observed in cows since they were first domesticated. In North America, about 40% of all dairy cows are infected in one or more quarters. The high incidence of mastitis can be attributed to deficiencies in management, poor milking procedures and equipment, inadequate housing, and the breeding of cows for ever-increasing yields of milk.

Figure 1. The Charm II test can detect antibiotics in milk down to levels as low as 10 parts per billion. That is like finding 10 kernels of corn in 3900 bushels of soybeans. (Courtesy of Dr. Stanley Charm, Malden, Massachusetts)

Types of Mastitis

Two kinds of mastitis can occur in your herd: subclinical and clinical. Subclinical mastitis may precede clinical mastitis. For every case of clinical mastitis you see in your herd, there are probably 15-40 cases of subclinical mastitis.

Subclinical Mastitis

A cow with subclinical mastitis doesn't show visible signs of infection. There are no clots or gargot in the milk. Signs of subclinical mastitis include an elevated somatic cell count, reduced output from the infected quarter and sometimes the quarter dries up. The udder and the milk usually appear normal in every other respect. Your vet can usually do a milk culture to identify the type of mastitis present. Subclinical mastitis also tends to be of longer duration and is more difficult to eliminate than clinical mastitis.

Clinical Mastitis

The signs of clinical mastitis are more obvious. The infection is frequently localized to one quarter which yields milk containing clots or gargot. Frequently the quarter is swollen and milk production is decreased. You can identify infected cows easily if you strip all your cows before you put on the milkers.

A major type of clinical mastitis is the *acute* systemic or toxemic mastitis. Affected cows secrete bad milk, lose their appetite, and appear depressed. Fever is common in less severe cases. In life-threatening cases, the rectal temperature may be depressed. A cow with systemic mastitis often has a rough haircoat and rapidly dehydrates. She may even appear lame if she has an extremely painful quarter with mastitis. The lameness is caused by a reluctance to put her leg down against the sore quarter. Diarrhea is also likely in toxemic mastitis.

Other changes occur in advanced cases of acute systemic mastitis. The cow may develop gangrene in the infected quarter if gangrene-causing organisms are present. Many of these cows are stressed and abort after recovery from the mastitis. They are often wobbly and staggery on their feet, and it's not unusual for them to go down and not get up.

With *chronic* clinical mastitis, the infection is usually localized to one quarter which secretes clots or decreased amounts of off-color milk. The cow isn't sick and seldom has a fever; she looks healthy.

How Mastitis Develops

Mastitis occurs because infectious organisms invade the udder via the teat canal. Occasionally mastitis shows up after a blow to the udder. The udder is an ideal environment for infections. It has a large blood supply and a lot of tissue where microorganisms can reproduce rapidly, especially in a cow that is just fresh. A fresh cow is more susceptible because her immune system is suppressed at calving. Because the udder frequently has edema, the blood flow may be compromised and it's difficult for antibodies and white blood cells to move into the infected area.

"I don't care how good she used to be! Look at those teats! She's a broken down old bag and I'd get rid of her!"

White blood cells are a major group of cells that fight infection; we call them somatic cells. Somatic cells in milk are made up of a number of different kinds of white blood cells, including secretory cells from the udder alveoli, and macrophages and neutrophils, which engulf bacteria for destruction. The somatic cells in the milk are a direct reflection of the white blood cells in the bloodstream. As an infection develops, macrophages and neutrophils flood the udder to destroy the infection (Figs 2, 3). Once the infection is eliminated, the macrophages and neutrophils are usually evacuated in the milk. The cow then sends other cells in to repair the damage to the udder. In many instances, the damage is serious and the cow can only repair the infection by filling the injured udder tissue with scar tissue. The result is a permanent loss of milk production from the scarred area of the udder.

Causes of Mastitis

Streptococcus Group

Bacteria of the species *Streptococcus* are a common cause of mastitis. This group consists of *Streptococcus agalactiae (Strep ag), Strep dysgalactiae, Strep uberis, Strep bovis* and *Enterococcus.*

Strep ag is a highly contagious organism that usually causes clinical mastitis with periodic flareups. It only lives in mammary tissue. *Strep ag* remains on the inner udder surfaces, where it causes local tissue damage. It cannot survive for very long outside on the teat skin or other external locations.

It spreads from quarter to quarter and from cow to cow by anything that has been contaminated with milk from infected quarters. Contaminated objects include milking machine teat cups or related milking equipment, udder rags or sponges, washwater, strip cups and the milker operator's hands. You can eradicate *Strep ag* from most herds because it is highly susceptible to antibiotics, such as penicillin. Since it doesn't invade deep into the udder tissue, it's readily destroyed. About 5% of herds free from *Strep ag* become reinfected each year because of careless milking procedures.

Strep dysgalactiae is quite similar to *Strep ag* but is not as common. The number of herds with infections of *Strep dysgalactiae* is lower. This organism is an environmental contaminant. Once it causes infection, it tends to move from a subclinical to a clinical stage sooner than some other organisms. Usually you can treat and eradicate the infection, especially if your cows don't have teat lesions.

The third organism in this *Streptococcus* group, *Strep uberis*, is quite different from the others. It's found nearly everywhere in a cow's environment, including the soil. *Strep uberis* can be recovered from manure, teat skin, the mouth and the rumen. Some *Strep uberis* infections are chronic and subclinical. *Strep uberis* is a very common cause of clinical mastitis in early lactation.

Figures 2 (top) and 3 (bottom). The circle is an abraded intramammary device (AIMD). The rod with the adaptor is used for installing an AIMD through the teat sphincter into a cow's teat cistern. The AIMD stimulates white blood cell production in the cistern. Using this device on a cow with a very low SCC gives greater protection against acute mastitis. (Courtesy of Dr. Max Paape, USDA, Beltsville, Maryland)

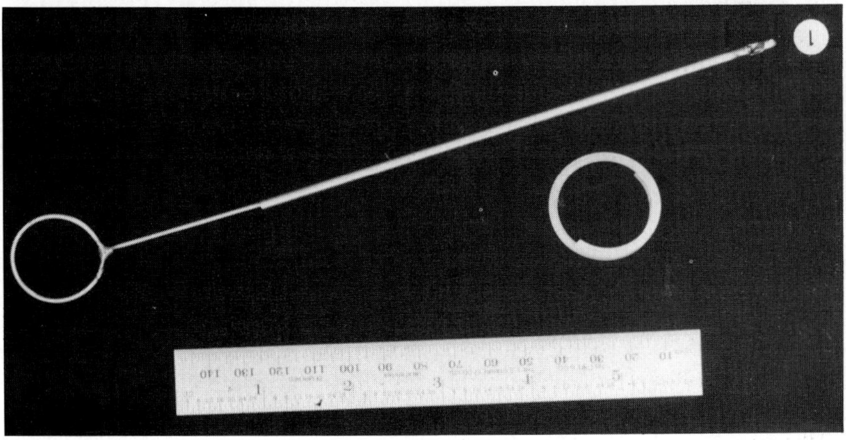

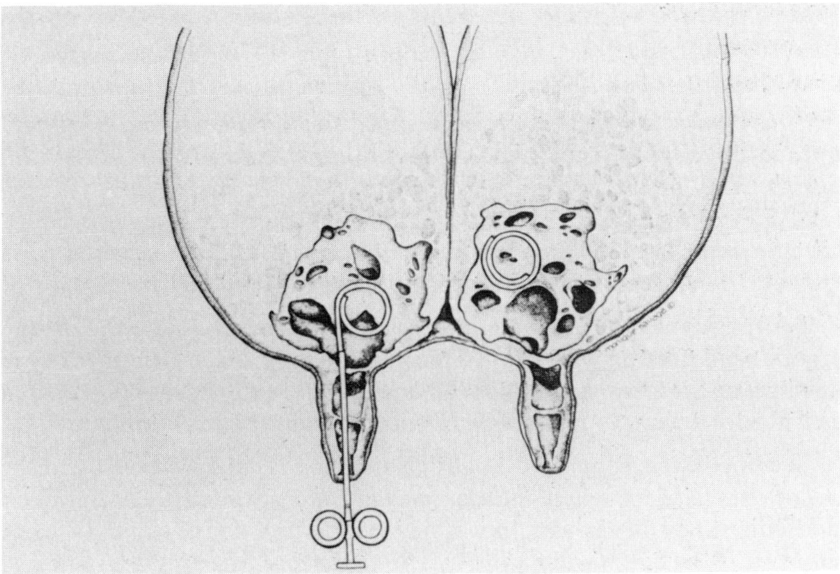

The incidence of new *Strep uberis* infections is also high during the second half of the dry period in herds that don't dry treat all cows. Many mastitis cases that occur after injury to the teat end are related to *Strep uberis*.

Strep bovis and *Enterococcus* are similar to *Strep uberis*. They are also commonly found in the environment and in manure. *Strep bovis* and *Enterococcus* are more resistant than other *Streps* to antibiotic therapy and eradication.

Streptococcus mastitis occurs in newly fresh cows as clinical mastitis. The milk looks normal but contains clots, pus or gargot. The cow often is not sick from the mastitis, but may have some tenderness in the quarter. Sometimes streptococcal mastitis is acute. Cows with acute mastitis have fever and get sick but usually their life isn't threatened.

Streptococcal mastitis is often subclinical and manifested only by a high somatic cell count. The bulk tank bacteria count is also higher than normal.

Staphylococcus Group

Staphylococcus aureus is very difficult to control in a herd once a number of cows become infected. It is quite contagious, but is not as serious a problem as *Strep agalactiae*. *Staph aureus* also inhabits the outside, as well as the inside, of the teat and udder skin. When a cow's udder has been injured from overmilking or improper milking machine function, the organism quickly infects the quarter. An infected quarter doesn't respond to treatment when a cow has *Staph* mastitis because the organism can penetrate deeply into the udder tissue where the antibiotic can't reach it.

Staph aureus is spread from infected to noninfected cows by contaminated milking equipment or the milker's hands. It can persist for months and is seldom eradicated very quickly. In many herds, it may take 2-3 years to eliminate *Staph* mastitis. The herd frequently requires such measures as teat dipping, dry cow therapy, milking infected cows last, and back flushing. Some infected cows must be culled.

Your vet may be able to have a vaccine made from the *Staph* organisms infecting your herd. This vaccine may reduce the severity of mastitis and lower the somatic cell count. But in my experience, *Staph* vaccinations seem to have little effect in preventing new infections.

In some herds, coagulase-negative strains of *Staphylococcus* are the major source of infection. These types of *Staph* are normally found on the skin. They often are the major population found in bulk tank samples. If the bulk tank has an elevated number of coagulase-negative *Staph*, it's likely that you need to do a better job preparing the udder for milking. You should check teat preparation practices, eliminate wet milking, and look for cows with rough, cracked teat skin. Elevated numbers of *Staph* are also common in herds that don't have their teats dipped. Almost all coagulase-negative *Staph* are eradicated by dry-cow therapy. Also, treating cows during lactation eliminates some of the infections from affected cows. With coagulase-negative *Staph*, there's a high rate of new infections during late lactation.

Acute *Staph* mastitis frequently produces systemic signs of illness. The recently fresh cow may be off feed, depressed, wobbly and toxic, and have a temperature as high as 107 F. Occasionally a sick cow goes down and can't get up. In the most severe forms of *Staph aureus* mastitis, the udder shows swelling, redness, heat, tenderness and, in some cases, gangrene. This is because of decreased blood flow to the infected quarter. The skin is cold, it becomes dark blue and tissue fluids ooze through it. The affected quarter feels cold and slimy. In the early stages, milk from that quarter may have clots and large flakes; then it becomes watery, blood-tinged and then dark red with blood. This form of *Staph* mastitis (gangrenous mastitis) doesn't respond well to therapy. The cow may recover only with intense therapy and under the direct supervision of a veterinarian. Even then, she may end up three-quartered (Fig 4).

Subclinical *Staph* mastitis has few typical signs other than low production in the herd. Some quarters may feel thick and meaty. Occasionally there are a few small flakes in milk from the infected quarter. A typical history of chronic *Staph* mastitis is that the cow is treated with mastitis tubes and immediately clears up, but the quarter

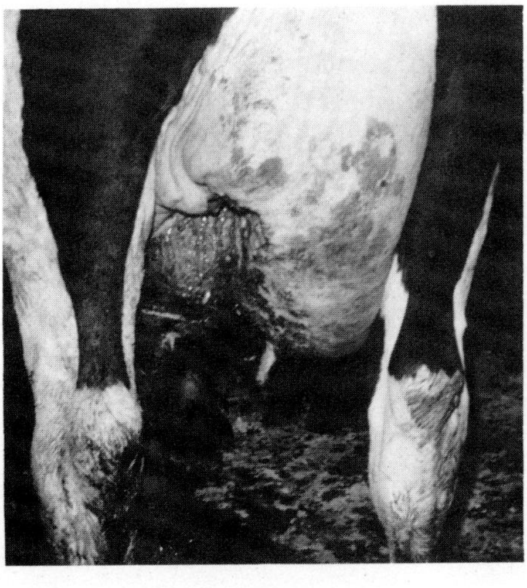

Figure 4. Though a cow may survive gangrenous mastitis, tissue damage is so extensive that the whole quarter drops off. This leaves a massive defect in the udder and limits the cow's usefulness.

soon flares up with mastitis again. Usually these chronic cases are closely associated with sublinical mastitis that has been in the cow for a long time. Subclinical *Staph* mastitis is extremely difficult to clear up because of its penetration into the udder tissue.

Coliforms

Coliform mastitis is caused by a number of bacteria that belong to one family. The most common member of the coliform family is *E coli*, an environmental contaminant. It's readily found in manure, buildings, barns, floors and cows. It is a normal inhabitant of the digestive tract of all animals.

Klebsiella is another member of the coliform group. It is an environmental contaminant that is common on bark and in sawdust from trees. *Klebsiella* causes mastitis very similar to that from *E coli*. The organism "rides into" the farm on loads of sawdust and shavings the dairyman uses for cattle bedding.

A third coliform that you may see on your vet's culture report is *Pseudomonas*. This bacterium is also an environmental contaminant found most often in manure.

Coliform infections cause acute systemic clinical mastitis. The cow is often very sick and off feed. In mildly affected cows, there is a high fever of 104-106 F. The affected quarter is painful and tender, and it secretes thin, watery milk containing a few clots of pus. The cow may be dehydrated, have diarrhea, dehydration and sunken eyes and in some instances she is toxic.

Coliforms can produce extremely toxic poisons that seriously affect the cow's metabolic processes. Cows with toxicity from coliform mastitis usually are wobbly and extremely depressed. They have diarrhea and may go down and be unable to get up. With this more severe form, the rectal temperature falls to below normal. The quarter secretions are thin and watery, and you can't milk much from the quarter. These coliform infections are life-threatening. Without intensive treatment, affected cows may die.

Chronic *E coli* mastitis is common in herds that have poor health. An affected cow has clots of pus in the quarter but she doesn't appear sick. If she's not fed well and properly cared for, her immune system won't be able to fight off this kind of infection.

Corynebacterium pyogenes

Corynebacterium pyogenes is a common pathogen that occasionally causes mastitis. This infection is usually low grade and without systemic involvement. Affected cows show clots of pus from a quarter and don't bounce back quickly, even with antibiotic therapy.

Leptospira

Leptospirosis can cause clinical mastitis, too. Five types of *Lepto* are common to dairy cattle (see Chapter 14). *Lepto* is usually involved with systemic disease in cows. One of the signs of this kind of mastitis is yellowy, ropy milk that yields no infectious organisms on routine culture. After you see the abnormal milk, the cow breaks with systemic signs of leptospirosis.

Fungi

Fungal types of mastitis are caused by *Candida* species and *Cryptococcus neoformans*. Most show up when cows are carelessly managed. Mastitic cows treated with antibiotics from multi-dose bottles are likely to get fungal mastitis. The multi-dose bottles become contaminated and the cow becomes infected with the fungus. Other cows get fungal mastitis from overzealous, long-term antibiotic therapy given in an attempt to clear up clinical mastitis. Fungal mastitis is generally unresponsive to antibiotic therapy.

Mycoplasma bovis

Mycoplasma bovis causes a type of chronic mastitis seen occasionally in large dairy herds, predominantly in the western United States. In the acute systemic form, the cow is sick, goes off feed and develops a fever. The infected quarter produces thin, watery milk and the quarter quickly dries up. The cow often dries up and may carry *Mycoplasma* organisms afterward. In the chronic form, the cow gives watery milk but she isn't sick. She continues to shed the organisms, and the rest of the herd becomes infected.

Why Do Cows Develop Mastitis?

Regardless of the type of mastitis you're confronted with, a cow picks up the infection from one or more of the following sources:

- Milker inflations
- Wash water
- Udder cloths
- Milkers' hands (see Chapter 17)
- Teat tubes, dilators, infusion tubes
- Herdmates and the environment

(For more details on better milking, see Chapter 17.)

Increased Susceptibility

Several factors increase a cow's susceptibility to mastitis. Some of these factors are associated with cow herself. You've seen a cow with a large pendulous udder become a chronic mastitis case. Any cow that leaks milk has a wide teat canal and is usually more susceptible. As these cows grow older, the teat canal tends to dilate and leak even worse. High-producing cows are stressed and always more prone to mastitis. Cows with long teats have a tendency to step on them.

Cows tend to become more susceptible to mastitis as they get older. You have probably also had more mastitis cases in cows that have just freshened and have udder edema. This is because a cow in the early stages of lactation has reduced immunity because of the stress of calving.

Heredity is also associated with mastitis. An iron-binding protein called lactoferrin is necessary for optimum immunity against mastitis. Some cows have a large amount of lactoferrin in their udders, while others don't. Some day, bull studs may provide information on this factor for sire selection.

Next, consider the cow's environment. Mastitis is more prevalent in wet weather. Other weather conditions also stress your cows and lead to increased mastitis. You don't have any control over the weather, but you can improve the ventilation in your barns to improve the air quality and reduce moisture levels stressing your cows. If you aren't actively using a fly-control program in the warm months of the year, your mastitis levels are also likely to get out of hand. Improper free-stall design is also a contributing factor to mastitis. Using organic material, such as straw, for bedding in your free stalls increases the incidence of environmental mastitis. You can cut the incidence by using sand or limestone, which reduces the number of bacteria in the stalls. Your herd size influences the number of cases of mastitis as well. If you have a large herd and fewer free stalls than cows, there is a greater risk of mastitis. Even the way you feed your herd can also make a cow susceptible to mastitis. Competing for bunk space is stressful and can predispose to mastitis.

A milking machine can carry infectious organisms from cow to cow and, when functioning improperly, push organisms up inside the teats. Milking machines can also reduce a cow's ability to resist infections, such as from *Staph aureus*.

Most dairymen aren't aware of the role that nutrition has on mastitis. Nutritional deficiencies or toxicities can increase a cow's susceptibility to mastitis. Deficiencies of trace minerals, such as selenium or copper, can cause mastitis. Mastitis outbreaks can occur because of toxicities.

Iodine toxicity suppresses the immune response to mastitis. Excessive dietary protein can raise the blood urea level and predispose to mastitis.

Many diseases also affect the cow's immune status and predispose the cow to mastitis. Johne's disease is a good example.

How is Mastitis Diagnosed?

You're familiar with diagnosing mastitis by stripping gargot from the quarter of a cow. Several other cowside tests can be used.

Physical Examination

Mastitis can be diagnosed by a physical exam that reveals an empty udder, unbalanced quarters, swelling in one quarter or other differences in firmness of a quarter. Stripping milk out on a strip plate or on the floor is also a good method of evaluating abnormalities. The strip cup has been eliminated in most herds because of the concern that it may harbor mastitis organisms. Stripping milk on the floor in parlors and using a hose to wash it away is an effective means of evaluating the milk while minimizing aerosol contamination with mastitis pathogens.

California Mastitis Test

Dairymen use the California mastitis test (CMT) to detect abnormal milk by observing the test's reaction to detect somatic cells in the milk. Use the CMT on cows that are ready for milking. Squirt a few streams of milk out before you conduct the test, and make a habit of recording the test results for use in the future. It's also important to recognize that the gel formed in the test in fragile and doesn't last long. Consequently, the CMT paddle should be rotated gently and the reaction read within 5-7 seconds. The samples are then scored as 0, trace, 1, 2 or 3, based upon the amount of gel formed. The scores are used to evaluate the number of somatic cells in the milk as follows: If there is no reaction and the treated milk remains thin, the somatic cell count is 100,000 or less. If the test shows a trace reaction, the count is usually 300,000. A CMT score of 1 indicates a somatic cell count of about 900,000. A CMT score of 2 indicates an average of 2,700,000 somatic cells, and a CMT score of 3 indicates a somatic cell count up to 8,000,000.

You may prefer to use a simpler scoring system: negative, suspect or positive. If you are having trouble deciding on the score, stick a finger into the treated milk and see how much gel hangs on the finger when you pull it out. Naturally, you should wash off your finger between cows. When done correctly, the test helps detect subclinical infections that often go unnoticed.

Bulk Tank Samples

Monitoring bulk tank milk is an important part of your dairy's production program. That information, in conjunction with the CMT and the somatic cell count, can provide an inexpensive means of monitoring milk quality and mastitis status on a regular basis. While culturing is valuable if you have a herd mastitis problem, it's often cumbersome and expensive for routine use. Bulk tank analysis can give you information on the extent of subclinical mastitis in your herd. It can also identify the infection, tell the level of your milking hygiene, and reveal how effectively your milking equipment is being washed. Bulk tank sampling can give you and your vet a significant edge on diagnosing mastitis problems before they come to the attention of the health department and milk processor.

Once you establish this bulk tank sampling as a regular procedure, you can identify problem cows and uncover the sites of milk residues that build up in your milking equipment. There generally are 5 tests that your vet will include in the routine evaluation of bulk tank milk: somatic cell count, coliform count, laboratory pasteurized count, standard plate count, and *Mycoplasma* cultures (Fig 5).

Probably the most valuable tool in bulk tank sampling is the somatic cell count (SCC). It can be measured either by a direct electronic count or with the Wisconsin mastitis test (WMT), a derivation of the CMT. The somatic cell count comes as close to giving you the status of mastitis in your herd as any one test.

Figure 5. The culture and sensitivity test is used to determine the cause of and treatment for an infection. On the left, agar is used to grow the organism causing the infection. On the right, the sensitivity test is used to determine which antibiotics are most likely to cure the infection.

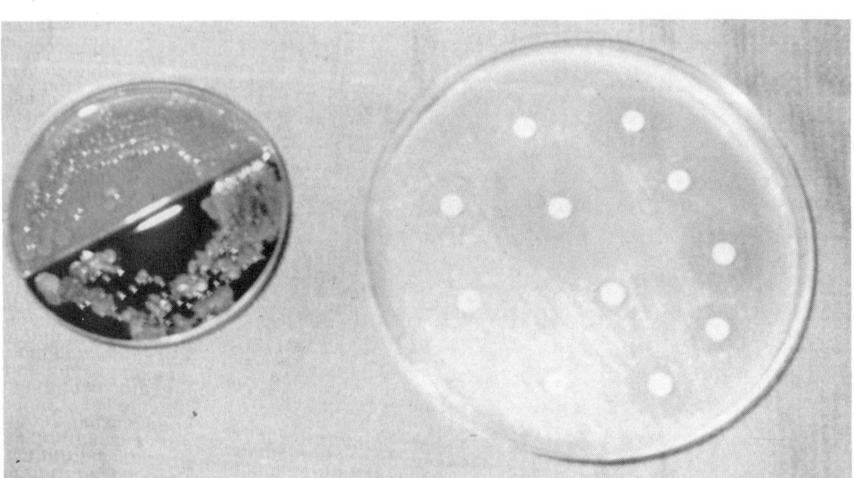

The coliform count is an excellent measure of milking hygiene and milking equipment sanitation. The laboratory pasteurized count reflects the degree of post-milking cleanup of your equipment. The standard plate count determines the number of bacteria in your milk. You can also use it to identify bacteria or other pathogens.

Mycoplasma cultures probably aren't necessary unless your herd is in California or the southwestern US. Many experts feel that *Mycoplasma* would be more often recognized as a problem across the country with more culturing for *Mycoplasma*. However, from the extensive *Mycoplasma* culturing I have done, I've not found this to be true.

Your vet or his lab personnel can perform most of these tests. They're basically set up as a modification of the standard tests that the milk processor and health department use. I encourage you to establish routine testing with your vet, because health departments and dairy processing plants are only concerned with numbers of organisms. You need to know not only the quantities of bacteria in your milk, but also the types as well. You and your vet then have clues as to the source of the mastitis and how it can be eliminated.

How is Mastitis Treated?

Mastitis can be treated in various ways, depending on the type and severity of infection. Numerous "snake oil" treatments have been touted all across the country. It's your job to sort these ineffective treatments out and ignore their marketers' claims.

Clinical Mastitis

Clinical mastitis cases can be mild without systemic involvement, or more severe, with systemic involvement. With a mild case, you may keep the quarter milked out and give oxytocin injections and not treat the cow at all. This ensures that none of the infected milk is held in the quarter. Strip the quarter every 2 hours. Many mild cases respond favorably to this treatment.

Intramammary infusions with mastitis tubes of penicillin or the newer synthetic antibiotics are likely to only clear up clinical signs of mastitis. The SCC is likely to stay elevated. Many people debate whether it pays off. Most evidence indicates that gains in milk production are only minimal. To my knowledge, there is only one report of cows with *Strep ag* being treated with mastitis tubes during lactation and returning to profitability. In most other reports, profitability was gained only by treating these cows during the dry period.

In acute (quick onset) cases of *Strep* and *Staph* mastitis with systemic involvement, more intensive treatment is necessary. I usually prescribe a broad-spectrum antibiotic. Large doses of aspirin or flunixin often are helpful. Flunixin should only be used at the direction of your vet because it is an extralabel usage drug..

Oxytocin should also be given with severe mastitis. It helps the udder empty itself of toxins. Frequent stripping of the infected quarter, in my opinion, is probably as important as the antibiotic therapy. In severe cases, your vet may recommend intravenous fluid therapy. Cortisone drugs, such as dexamethasone, give a temporary "lift" to the cow so that she eats better, and also reduce udder swelling. Dexamethasone, however, may suppress the immune response.

Many cows die of coliform mastitis unless you provide intensive care. A top priority is frequent stripping of the infected quarter; keep it empty. Stripping and aspirin-like drugs should be the backbone of therapy. Always consult your vet about the best treatments and the milk withdrawal and residue times. Your vet may give intravenous fluids to a cow with coliform mastitis. It can mean the difference between life and death.

Some severely affected cows may be down and unable to get up. In these instances, your vet may give slow intravenous calcium. It's usually safer to dilute it out in several gallons of the intravenous fluid. Giving intravenous calcium alone may cause the cow to go into cardiac arrest.

Give severely affected cows intensive nursing care and rely on veterinary supervision. I only prescribe corticosteroids, such as dexamethasone, in cases of shock. If you've ever lost a cow to this kind of mastitis, you'll be happy to know that researchers are developing some new therapies.

Subclinical Mastitis

Subclinical mastitis is never life-threatening. However, cows with subclinical mastitis may have to go to slaughter unless you give proper treatment. I usually tell my clients that it's important to treat subclinical cases, but the economics of treating cows during lactation are variable. You only get payback when your treatment drastically drops the somatic cell count.

Let's say that your cow is milking and you've decided to go ahead and treat her mastitis. You have several choices. Many vets treat on the basis of a high SCC and cultures (Fig 5). They usually recommend intramammary infusion. A "blitz" treatment of penicillin may eliminate *Strep ag* mastitis, but cure rates on other types can be poor with this approach.

The infection may be eliminated or reduced but the SCC usually remain high for some time.

The dry period is the best time to treat subclinical mastitis. There is a reduced incidence of new infections at that time. The cure rate is considerably higher because of the longer effect of dry-cow mastitis tubes. Damaged tissue has time to regenerate before the cow starts milking again. Dry-cow treatment also reduces the number of clinical cases of mastitis at freshening. In addition, you don't have the risk of antibiotic residues in milk. But keep in mind that you must observe withdrawal times on treated cows you decide to slaughter.

You must always use approved dry-cow antibiotic infusion tubes in a slow-release base for dry-cow therapy. *Do not use products labeled for lactating cows*. Give the dry-cow mastitis therapy after the final milking. The antibiotic becomes dispersed throughout the quarter. Treat all cows you're drying up. When you treat selectively, you may clear up existing infections but you can't prevent new ones from developing.

Response to Treatment

Regardless of whether you treat clinical or subclinical mastitis, many dairymen are disappointed that the somatic cell count fails to drop as soon as the milk appears normal. Be realistic; it may take 2 weeks to several months for the somatic cell count to fall.

It's also important to observe drug withdrawal times on antibiotics and be sure that you've discarded milk from all quarters. Don't assume that you can treat one quarter, discard the milk from that quarter, and not get caught with tank residues when putting milk from the other

"They're going to shut down the FDA. They've been causing cancer in rats."

untreated quarters in the tank. This is absolutely not true. You may find it useful to buy a Delvo Test Kit to check the milk from treated cows for drug residues. There are also other new milk residue test kits being developed.

When cows have active infections, do not feed their milk to the calves. Once you've administered antibiotic therapy and the milk appears normal, you can feed it to calves. While some controversy still exists about this, none of the controlled research indicates that there's a problem with feeding antibiotic milk to calves unable to suck each other. But do not feed mastitic milk from treated cows to bob veal (calves that are going to the market shortly after birth). These bob veal will have antibiotic residues at the slaughterhouse.

A Mastitis Control Program

A good mastitis control program has 5 major facets:

1. *Correctly use properly functioning milking machines and properly prepare udders.*

- Attach teat cups after you've thoroughly cleaned and dried the teats and bottom of the udder.
- Provide stable vacuum of 11-13 inches of mercury at the short milk tube during peak milk flow.
- Avoid squawking or slipping of teat cup liners during milking.
- Shut off the vacuum to the claw before you remove teat cups.

2. *Dip teats at milking.* Mastitis organisms are likely to spread at milking time, even under the best management conditions. To destroy organisms before milking, use a premilking teat dip. To destroy the organisms remaining on the teats at the end of the milking, use a postmilking teat dip. Pre-dip cows with a teat dip instead of washing with a hose. Then wipe and massage the teat dip off before attaching the milker. Occasionally, cows may be so filthy that you need to wash them before you use the pre-dip on the teats. Dry them, then put on the milking machine, milk, then dip the teats again.

Extensive research has been done on most teat dips on the market. Most teat dip products reduce new infections by 50-80% when you use them in conjunction with other components of mastitis control. Teat dip products are highly effective when they contain one of the following: 0.1-1.0% iodine, 0.2% chlorhexidine, 2% bromine, 0.5 quaternary ammonium, or 2% linear dodecyl benzene sulfonic acid. Use only 0.5% iodine as a pre-dip so you don't risk contaminating the milk, or another type of dip expressly marketed for pre-dipping. Many dairymen use 4%

sodium hypochlorite as teat dip. While considered by many to be very effective, it is not an approved teat dip.

3. *When mastitis has become a problem, treat clinical cases properly.* The factors of correct treatment include:

- Treat cows for one additional day after you've noticed a favorable response to treatment.
- Treat cows for a minimum of 3 days even if they've cleared up early.
- Observe correct drug withdrawal times.
- Protect newly affected cows, especially fresh cows, from extreme weather conditions. Alternate applications of hot and cold water to the udder.
- Massage and strip out the milk and debris in the hot, acute cases of mastitis.
- Give oxytocin at 1- to 8-hour intervals, with followup stripping of the infected quarter. In severe cases, call your vet.

4. *Treat each cow as you turn her dry.* Use a specially formulated, commercially available dry-cow treatment, and treat all quarters. This covers all infected cows. In addition, you build in prevention against new dry-cow infections. With this regular treatment, you rarely need to culture milk samples. Other advantages of dry-cow treatment are:

The cure rate during the dry period is about twice as high during lactation. Damaged tissue may regenerate before freshening so that the cow that went dry with a blank quarter from a previous infection may very well start secreting milk in that quarter at the next freshening. Dry-cow treatment reduces new cases of clinical mastitis at freshening. Finally, by treating cows that are turned dry, salable milk is not contaminated with drug residues.

5. *Cull cows with chronic mastitis.* Cows that specifically fail to respond favorably to treatment or continue to flare up with repeated clinical mastitis should be culled. Remember that their continued presence in the herd may result in spread of infection to other cows.

I don't include vaccination as one of the 5 factors in controlling mastitis because most mastitis vaccines don't work. The one exception may be *Staph*. Your vet can develop a *Staph* vaccine from organisms identified in the herd. It may not prevent new infections but it may help reduce SCC in previously infected cows.

A Look to the Future

Continuing research in mastitis control uncovers some new truths. We know now that SCC will be highest in herds that are deficient in vitamin E and selenium. So that's a place to start looking if you have chronic mastitis in your herd. Treatment response will improve if you get vitamin E and selenium up to optimum dietary levels (see Chapter 19).

Another new development you may have heard about is the abraded intramammary device (Figs 2, 3). Cooperative Israeli and USDA research indicated initially that putting a roughened foreign body in the teat cistern improves a cow's immunity to new infections. This device was nothing more than a tiny rough piece of plastic inserted through the teat and left in the quarter. The theory was that the foreign body would stimulate the immune system, reducing the number of new infections in these cows. Recent research on these devices reveals initially that they do stimulate the immune response; however, later they break up in the quarter. Upon deterioration of the plastic, the udder cells develop a severe tissue reaction.

New genetically engineered compounds called cytokines are being evaluated to improve the immune response in the udder. These include colony stimulating factor (CSF), granulocyte-stimulating factor (GSF), interleukin-1 and interleukin-2, interferon and r-bovine interferon. It is also noteworthy that bovine somatotropin (BST) stimulates the immune response in the udder. Proposed uses of these compounds are in dry cow preparations or in treatment of severe mastitis.

16

The Milking Machine and Udder Health

The milking machine is the most important piece of equipment you own. This machine allows you to harvest your crop. It can make you or break you. I have clients that overhaul tractors, update harvesting and planting equipment, and keep their pickups polished. Yet, they never look at the milking equipment they use twice daily, every day, 365 days a year. The milking machine is used more hours a year than any other piece of farm equipment (Fig 1).

During this day-in and day-out use, the milking machine has a number of key components that must function together for safe and efficient harvesting of milk. The system is only as good as its weakest link. You should understand the function and maintenance of the various components of your equipment. Your understanding must be thorough enough so you can perform routine maintenance. And, you must be able to recognize if your equipment is malfunctioning — before it injures your cows or fails all together. Most important, your equipment should be updated when improvements are introduced that will increase milking efficiency.

Figure 1. Though the milking machine is used twice a day every day, it is often the most neglected piece of equipment on the farm. A poorly maintained air controller like this one will cause fluctuating vacuum and subsequent mastitis.

Your milking machine comes in contact with a cow's udder twice a day for 10 months of the year. It must caress a cow's udder rather than just suck the milk out of it. The udder is a delicate, easily contaminated organ that can yield large profits if it is handled gently. Mishandled, it becomes contaminated and infected, then less profitable.

The Milking System

Vacuum Pump and Vacuum Controller

You need a basic understanding of the mechanics of milkers. If you already know the basics, jump ahead to the next section.

Vacuum or suction is the principle on which all milking systems operate. The vacuum is created by removing air from a sealed container, namely the liner, claw and milk hoses. Two factors must be considered when discussing standards for vacuum in your system: negative pressure and air flow.

A Lot of Hot Air

I'll never forget going to a farm with a high SCC years ago. This dairy was many miles from my office and I hadn't previously known the dairyman or his equipment dealer. In checking the milking system over, I found it to be woefully deficient in pump capacity. I measured air flow and found the cubic feet of air flow per minute (CFM) to be 25% of adequate levels. The dealer had been quiet during my analysis. Finally, he said that his meter didn't agree with mine, and he offered to show me the measurements. He shut off the vacuum controller and put a pressure gauge on the line at the pump. Quickly turning on the pump before I could get a word in edgewise, he let the pressure rise while measuring it on his gauge. When the negative pressure got to 23 inches of mercury, there was a loud bang. It scared all of us. Quickly regaining his composure, the dealer said, "I measured 23 inches." Well, we had collapsed the balance tank during this demonstration. Fortunately, the balance tank could easily be circumvented for that night's milking. What the dealer hadn't realized was that I was measuring air flow while he was measuring pressure. I am happy to report today that the milking machine dealers that I work with have a much better background for their service work.

Unless you have one of the old dual pump milking systems, your pump creates negative pressure in 2 subsystems. The milk line removes and transports milk. The pulsator line creates cycles of vacuum to add safety for the cow's udder. Your vacuum system must provide equal amounts of vacuum to both subsystems. Nearly all milker systems use one vacuum source to provide vacuum for both subsystems. There are a couple of exceptions. The Zero system may provide one pump for transporting milk and another for pulsation and milk extraction from the cow. In earlier years, Surge had a system called the "Tonganoxie" system. With this milker, one pump provided vacuum to the milk line and claw, while the other provided vacuum to the pulsation lines and pulsators. Surge has replaced nearly all of these systems with improved vacuum systems having one vacuum source.

Vacuum pumps are rated in CFMs (cubic feet per minute). While electric motor horsepower gives you a rough estimate of pump size, horsepower is not a measure of pump capacity. Roughly, 1 HP is equal to 10-12 CFM.

Vacuum pump ratings must have 2 interrelated specifications. They are CFM at a specified negative air pressure. Negative air pressure is measured with a mercury manometer that compares the negative pressure against atmospheric pressure. For vacuum pump pressure, "inches" of mercury refers to the inches of mercury that a manometer is deflected away from atmospheric level.

The CFM is measured with a flow meter. The CFM unit is measured in either of 2 unit systems. Both are called CFMs, but the New Zealand measurement is just half of the American standard developed by the

"I told you to cut DOWN on the vacuum!"

American Society of Mechanical Engineers (ASME). You should remember which units your dealer is using when discussing your pump capacity. Just remember that the New Zealand standard requires twice as many CFMs to do the same job as the American standard. Some companies use one standard, while others use the opposite.

Your pump size is determined by the air flow demand of the various components of your milking system. Many factors are entered into its determination.

- Negative air pressure at which the system will operate.
- Number of milker units and type of pulsation.
- Length and diameter of the pulsator and milk lines.
- Milk meters in the system or if there is a milk releaser.
- Vacuum-operated takeoff systems.
- Number of elbows or tees in the lines.
- Any factor that will restrict milk or air flow.
- Safety factors are included because pumps gradually wear out. The pump is often slightly oversized to compensate for wear.

Milking machine manufacturers have established guidelines, called 3-A standards, for their installations. These minimum guidelines establish safety levels for their equipment. While manufacturers have conscientiously calculated values for your system, you can use a rule of thumb to get an estimate of the size of pump required for your system. Generally, a pump that provides 10 CFM (ASME) is adequate for each milker unit without automatic takeoffs.

The vacuum controller (sometimes called the vacuum regulator) is a very important piece of safety equipment in your system. The controller is a spring-loaded or weighted diaphragm device that continually bleeds air into the system. This maintains a constant level of vacuum at the cow's udder. The controller must be rated to handle at least as much air flow as your vacuum pump. Fluctuations should not exceed 1 inch on a mercury manometer during operation. The controller should usually be located close to the vacuum pump, in a location that is relatively dust and dirt free (Fig 1). It should also be protected from the weather.

Vacuum lines should be installed with the same concern for safety as you had when you installed the pump and controller. Vacuum lines should make a complete circuit with no dead ends in the lines. Keep elbows and tees to a minimum because these create restrictions in the system. Street els are also a source of restriction.

See that the milk receiver has no restrictions of either air flow or milk flow. It serves to separate air from the milk so that vacuum can be maintained while the milk flows to the tank. Some very old systems still use a vacuum releaser instead of a receiver jar. Most of these systems are antiquated and should be replaced. An exception is one in which a vacuum releaser is provided for each cow to measure milk and control the amount of feed available to her in the parlor. This type of releaser is rarely used any more.

The vacuum system should be checked regularly to ensure proper operation. It causes malfunction when it becomes clogged with milk solids, dust, dirt and cow hair.

Maintenance is Important

Once I was called to look at Jack's herd because he had a high somatic cell count. He had good local veterinary service but he wanted an outside expert. (The expert is the guy that is called in from more than 50 miles away.) I remember him for his eccentric character more than the fact that he never paid his bill. Jack milked in stanchions with DeLaval units. He couldn't understand why he wasn't able to do an adequate job milking his cows. "Things just aren't working right," he said. He had a high SCC and, true to the usual form, he hadn't called me until he was ready to get shut off from selling Grade-A milk.

After I evaluated his milking equipment, it was apparent that the design of the equipment was really excellent. The pulsators worked efficiently. The vacuum pump also worked at specification. The regulator was not sticking, and Jack was doing a decent job of udder preparation. I suggested to Jack that we take his pulsator line apart and look at it. When he opened the tee in the end of the line, I had never seen a line as thoroughly plugged as Jack's. The 2-inch galvanized line only had an opening just big enough to stick a screw driver down the center of it. The rest of it was plugged full of milk solids, rust and dirt. This was why Jack's milker wasn't functioning. Jack had not looked at his pulsator line in over 2 years. I told Jack his milker system had hardening of the arteries. He didn't appreciate my humor.

The Milking Unit, Pulsators and Pulsation

The milking unit has 2 functions. It delivers the vacuum to the teat end for milk removal and then collects and channels the milk toward the tank.

A number of components of the milking unit work together for correct function of the unit. The teat cups, of course, are attached to the teats. They are made of a rigid metal or plastic outer shell and a flexible inner liner. An inner liner manufactured from a mold lasts longer than other types. Molded liners, however, do not provide the quality massage provided by other types of liners. Maintaining consistent quality in the manufacturing process is also sometimes a problem.

A stretch liner, in contrast, is manufactured as a tube. By means of a ring and retainer, it is stretched to conform to the shell. The stretch liner massages the teat more effectively than the molded liner but doesn't last as long. The quality of the product when manufactured is more consistent than the molded liner.

In some areas of the country, square molded liners are available. Though they vary in quality, massage with the square liner is excellent. Sometimes these liners are of concern to health inspectors because they may not be properly cleaned.

Narrow-bore liners are more desirable than wide-bore liners, which are antiquated. Wide-bore liners attach to the udder rather than sealing on the teats.

The claw is the most common type of collection unit. The teat cups usually attach to the claw by the short milk tubes. The claw has 2 functions. It is a small reservoir to collect the milk before it is transported from the cow and it provides a reserve of air flow for milk extraction. The claw system has the advantage of being light weight and easily maneuvered onto the cow. The disadvantage of the claw is that it can restrict milk flow away from the cow.

The second type of collection unit connected to the short milk tubes is the suspension system. Few are used nowadays. This system has the advantage of putting the milk directly into a bucket or breaker cup without any restrictions. It also provides a large vacuum reserve close to the teats. This system has a number of disadvantages. Often the pulsation depends on the vacuum to operate, so pulsation is susceptible to vacuum fluctuations. While a vacuum reserve is close by, these systems usually have a small vacuum pump design. There is also considerable effort involved in pulling a heavy bucket full of milk or breaker cup out from under the cow. Cows require a strap over their back to support the unit. In more recent times, the suspension was modified with a strap and a claw unit.

Automatic takeoff (ATO) units are often used with the claw type of milker. This is a separate system added to the milker. It has a sensor to determine when milk flow has stopped so the milker can be removed. The automatic sensing unit triggers the vacuum to be turned off and released. At this point, a mechanical arm or cord removes the milker. Most systems remove all of the teat cups at once. Years ago, a quarter

Figure 2. Automatic takeoffs are an excellent labor-saving device. (Courtesy of Germania Dairy Automation)

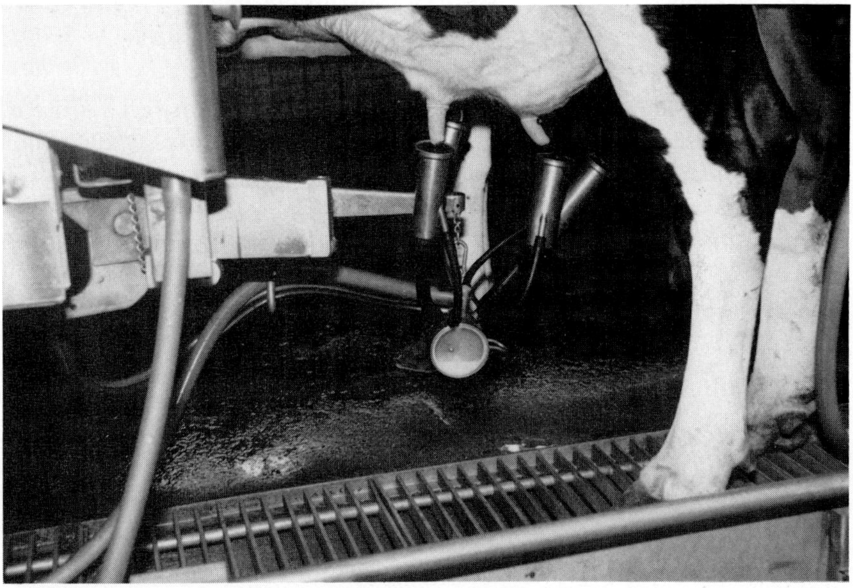

take-off unit was available. While considerable engineering went into this idea, it never became very popular. One of the major reasons was the greater cost of the quarter takeoff system (Fig 2).

Some systems reduce the vacuum and change the pulsation rate rather than taking the unit off. Unless they are in the hands of a mechanically inclined conscientious operator, they don't work well. Cows usually start letting their milk down immediately upon attachment of the unit, so units that change pulsation rates are often too slow in responding.

There are a number of advantages to an ATO system. Most important, they reduce labor needs in large herds. They also eliminate much of the problem with overmilking. If you are considering ATO systems, there are some problems to be overcome. First, they cost a bundle and they require regular maintenance and updating. They also allow less opportunity to provide individual udder care for problem cows. If a cow milks slowly in one quarter, the other quarters may be overmilked or the slow quarter may not be milked out properly. This requires that the operator milk the slow quarter out by hand.

Air vents are available for most milker units. They admit air to the short milk tube just below the shell. Air admitted to the short milk tube breaks the column of milk. This is supposed to increase the velocity of milk being carried away from the teat and thereby reduce flooding of the short milk tube. It also eliminates solid columns of milk in the line so there isn't a hydraulic effect to vacuum fluctuations. If a manufacturer doesn't provide air vents in the short milk tube, they always have one in the claw. It is thought that if the vents are in the milk tubes rather than the claw, flooding is lessened at the teat end.

It has been questioned whether air vents are necessary on a system. They often accentuate problems if the vacuum pump is marginal. If air vents are located at both the claw and the short milk tube, you may collect foaming and rancid milk. Plug the claw vent if air vents are used in the short milk tubes. An advantage of air vents on each milk tube is that if one is plugged, air is still admitted in the other 3. A design currently in research allows air to be admitted in the liner.

Milking Unit Function

The cow's teats are opened by the vacuum from the teat cup combined with internal pressure from milk letdown. The pulsator opens and closes the liner. Contrary to what many think, milk flow occurs while the liner is open rather than closed. The closed phase for the liner is to allow the cow's circulation to move blood through the teat end. Without pulsation, blood would collect in the teat end. This causes irritation to the teats and subsequently a high SCC.

The teat end vacuum must reach zero for a few milliseconds before the pulsator closes. This opens the liner and changes the vacuum back to the milk phase. The liner's opening and closing provides massage for the teat end and helps to maintain a healthy teat.

The pulsator closes the liner by a port opening in the pulsator, which allows air into the pulsator hose, the short pulsator tube and between the liner and shell. The liner is opened by the pulsator port's closing, which allows vacuum to be restored between the liner and the shell.

Several engineering designs are commonly used to pulsate the vacuum in the teat cup. Some systems use a centrally controlled, electrically operated pulsator to let air into the system. This method reliably controls the pulsator, but stray voltage can be a problem if this design malfunctions.

Such is not the case with a second design called air-operated central-control pulsators. A third method of pulsation is the self-activated pneumatic pulsator on each claw or bucket. This design is simple in operation but it is susceptible to fluctuations in vacuum. An operator can regulate the pulsator by making adjustments during the milking process. Consequently, the cow is at the mercy of the whims of the operator.

You should be aware of certain factors regarding pulsators. *The vacuum level between the shell and liner should be equal to the vacuum in the liner on the teat end*. The pulsation ratio should be about 60:40. This means that vacuum should be in the pulsator 60% of the time, and the pulsator should be open and admitting air 40% of the time. Because of the time involved in opening and closing the liner and the vacuum needed to restore milk flow, the milk:rest ratio on a correctly operating system will be slightly higher than the pulsation ratio. Your pulsation ratio should match manufacturer specifications. The pulsation rate is the number of times that the pulsator opens and closes in a minute. Usually, the rate shouldn't exceed 60/minute.

One manufacturer has often installed equipment with a high pulsation ratio (70:30). Though cows milk faster at this ratio, this company, in the past, used a small vacuum pump and high line. This combination of equipment often resulted in a higher SCC. Remember that a high pulsation ratio can become a significant problem if you have inadequate vacuum reserves. This combination of events creates a very high milk: rest ratio, which allows little time for recovery of blood flow in the teat.

Testing Milking Unit Function

There are no simple tests to check the performance of a milking unit. A number of clues can point toward poor unit function. Observe how the teat cup attaches to the teat. Vacuum should only be applied to the end of the teat. This subjects a smaller area of skin to impaired blood flow in the teat. By minimizing the area affected by the vacuum, the teat cistern remains open for milk flow into the teat canal.

Liner diameter and tension affect how the teat cup hangs on the teat. Vacuum draws the teat cup up toward the base of the udder as the cow is milked. The teats also contract, so the net effect is that the teat cup creeps toward the base of the udder. As the teat cup creeps upward, more skin is drawn into the opening, gradually shutting off milk flow into the teat canal. The teat canal isn't filled with milk as the liner opens, so vacuum in the liner is transferred into the empty teat. No milk is available to cushion the teat from the direct effect of vacuum on the tissue. This stresses the teat tissue. Without milk, there also isn't any milk flow to prevent the entrance of bacteria into the teat. The result is bacterial invasion of the teat canal. Many milker manufacturers put lead weights in the claw assembly to help prevent the unit from creeping up the udder.

The flexibility of the liner also affects unit function on the teat. You have probably observed that the shape and size of the liner mouthpiece also affect performance. Various types of liners are available on the market. If you aren't satisfied with the type you are using, you may need to experiment to find the liner that works best on your cows. Ask your local serviceman for advice about different liners.

The capacity of the claw has a direct influence on how each unit performs. Generally, larger claw assemblies are a better choice to prevent flooding your system. Look at new-design claws available from your serviceman to decide if you need to update. If your claws are considerably smaller than the newer styles, you should consider updating.

The diameter and length of the short milk tubes also influence unit performance. Follow your serviceman's advice. Air vents have a direct influence on unit efficiency. Air vents in the claw, while very functional for transport in the long milk hose, do not enhance milk transport away from the teat ends. Air vents installed on each quarter as close to the teat end as possible appear to give the best milk flow.

Newer designs in claws prevent the rebound or impact of milk and air off the floor of the claw and up into the opposite milk tube. Corrugations on the inside surface of the claw eliminate the rebound (Fig 3). Of course, the amount of teat cup slippage allowed by the operator has a lot to do with whether you need this design. Unstable vacuum also allows the teat end to be bombarded by droplets laden with more organisms, especially when there is teat cup slippage.

Figure 3. When air leaks into squawking teat cups, milk, bacteria and air shoot down the short milk tube and rebound in the claw up into the other teat cups. Bacteria driven against the teat ends may cause mastitis.

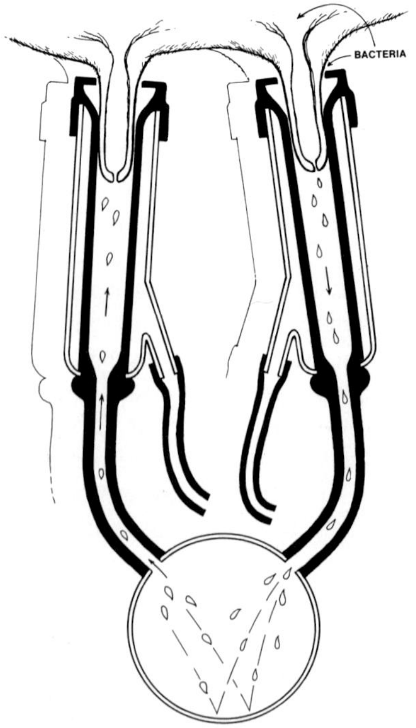

A critical factor for proper milker unit operation is the vacuum created by the vacuum pump. Without proper pump function, other aspects of the unit are relatively unimportant. You can best determine if your pump is sufficient by having your serviceman measure CFMs. You should have 10 CFM/unit (ASME); air flow of less than 7 CFM is cause for concern.

Unit function is also affected by several factors you can't control. Most important of these is the shape of the cow's teats. Long, thin teats tend to allow a unit to creep upward. Cows with pendulous udders and unbalanced quarters also don't milk out evenly (Fig 4).

The Milking Machine and Udder Health

There is a very direct link between milking machine function and the incidence of mastitis. The milking machine is usually responsible for carrying mastitis-causing organisms to cows. Infectious organisms are always present in your cows' environment. The milking machine easily becomes contaminated with these organisms and spreads them from cow to cow.

Your goal should be to prevent the spread of organisms by the machine. You can do this by regularly replacing cracked and worn liners, hoses and gaskets. Wash and sanitize the milker after every milking. Milk infected cows last if you can manage it. Practice good sanitation and good milking procedures (see Chapter 15).

Your milking machine impairs a cow's ability to resist infections. The normal anatomy of a cow's teat is designed to resist penetration by

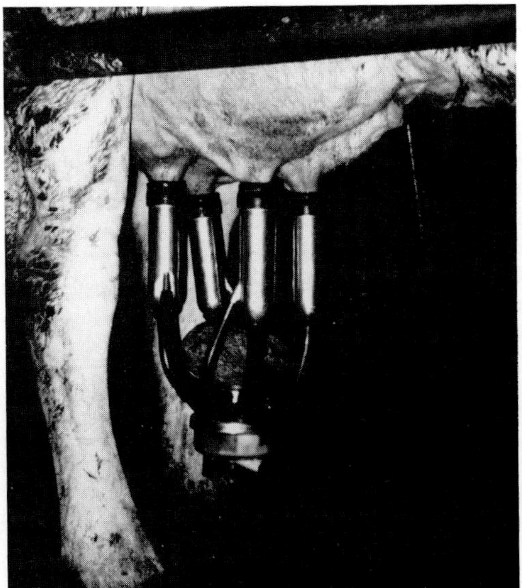

Figure 4. Some herdsmen put a large weight on the claw to improve milk-out time of hard-milking cows. Most of these operators never realize that their milker performance is likely substandard. The milking machine should be updated to correct the problem of insufficient air flow.

organisms. The teat has natural resistance to being opened. The teat canal has a waxy keratin lining with a germicidal property. A milking machine tends to break down this natural resistance to infection. This is especially true when you overmilk a cow. Excessive vacuum greatly reduces a cow's ability to resist infection. Poorly designed or inappropriate liners can also increase susceptibility to infection.

The signs of damage to the teat canal can be readily spotted while you are milking. Because of teat pain, cows are likely to be "kicky." If you check the teat ends, the teat sphincter may be turned outward. Some cows may have a bluish cast to the teats after the unit comes off. This means that blood circulation to the teat is impaired. Other cows may have tiny blood spots like freckles on the teat skin. Older cows often show scarred, misshapen teats.

In addition to the milker's traumatizing the teats, it readily spreads organisms into the teats (Fig 5). This usually happens near the end of milk flow while the milker is still attached. The flow is at a low level so it doesn't wash the organisms out that are injected toward the teat end.

Detecting and Solving Vacuum Problems

One of the most common problems I observe is caused by inadequate vacuum airflow to the claw. You can detect this if you know what to look for. The cows tend to milk out slowly, milker unit dropoff is a problem and the SCC is higher than normal. Production always suffers

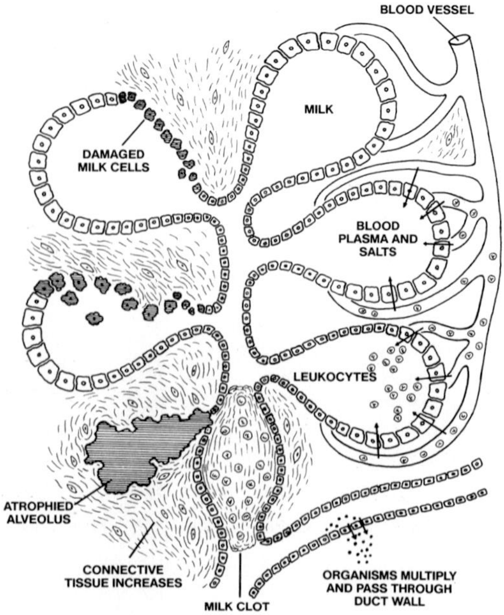

Figure 5. Most cases of mastitis begin with entrance of mastitis-causing bacteria into the udder through the streak canal of the teat. This microscopic view shows inflammation from entrance of bacteria. Connective or scar tissue replaces milk-producing alveoli after these bacterial attacks.

from these conditions. If vacuum is really low, the liners don't open fully. This can be best detected by graphing milker performance.

Once you have determined that you have vacuum problems, start your investigation. First, have your serviceman check vacuum pump performance. You may have inadequate capacity or the pump could be worn. Sometimes, it is nothing more than slipping belts. There may be leaks in the system or, occasionally, the vacuum lines are too small. This becomes more serious if the lines have numerous turns and fittings.

If pulsator lines become filled with debris from overflowing milk traps or cracked inflations, this can also restrict air flow. Inadequate air flow can occur if pulsators are worn, dirty and malfunctioning. Vacuum can be influenced by a worn, sticky, small vacuum controller. To cut costs, some producers use cheap inflations and shells that are ill-designed for their system. Sometimes, an improperly installed vacuum controller, pump or lines causes inadequate air flow.

Excessive vacuum to the claw can be devastating to udder health. The first signs are faster milking and teat cups riding up on the udder. The cows tend to have incomplete milkout and become kicky. Teats gradually begin to show bruising and discoloration. Teat tissue gets crowded into the mouthpiece and shuts off milk flow into the teat canal. The SCC then gradually increases.

Excessive vacuum generally is related to either a small or sticking vacuum controller. Incorrect inflations may also stimulate excessive vacuum levels. Clogged air vents cause excessive vacuum on the teats.

Cyclic fluctuations in vacuum level have a detrimental influence on udder health. Cyclic fluctuations occur in sequence with the action of the pulsators. Signs are vague, but the SCC consistently creeps upward.

Vacuum fluctuations usually occur because of restrictions in air flow. The claw may be too small or the short milk tubes may have a narrow diameter. Small vacuum lines or small air vents also cause cyclic fluctuations. Sometimes vacuum fluctuations can be caused by overly snappy action of the pulsators. Cyclic fluctuations readily occur when the reserve air capacity is too small for the system. Don't forget to first check the vacuum pump belts before tearing into the system to look for a major defect.

Random fluctuations in vacuum may occur during milking. The only sign is a high SCC. Irregular fluctuations, in my experience, are most common when the milker is attached and detached carelessly. Too much air is allowed into the system. This is usually compounded by a sticky vacuum controller and inadequate vacuum capacity. Small vacuum lines also enhance the effects of these fluctuations. In any case, the root of the problem is allowing too much air into the system without a quick response from reserve air to correct the potential fluctuation.

If you spot a potential milker malfunction, arrange for your serviceman to visit. He will want to evaluate your system thoroughly. The CFMs must be measured at the vacuum pump. This is done with an air flow meter. You are in pretty good shape if you have 10-12 CFM/unit (ASME), not considering automatic takeoffs.

The reserve air flow is determined by using the air flow meter on the far end of the system, with the pulsators and controller working. Reserve air should be 50% or more of vacuum pump measurements.

After determining reserve air, record measurements for each unit while milking the cow. Ideally, this is done with a dual channel recorder. A dual channel recorder can measure pulsator vacuum simultaneous to milk vacuum measurements. One channel is connected to the short pulsator tube and the other is connected to the short milk tube. The recording is started as the unit is attached to the cow (Fig 6). Your serviceman can evaluate the recordings after the milker is detached. He will check the tape for pulsator fluctuations or the milking vacuum's fluctuating independently of the pulsators. Sometimes abnormal fluctuation on the milk line is related to the pulsators. Some companies have their servicemen use single-channel recorders, measuring only pulsator or milk line vacuum. While this isn't detrimental, your serviceman can better tell if the fluctuations are cyclic by taking measurements simultaneously with a dual-channel recorder.

Figure 6. In recent years, vacuum monitors have been developed to measure vacuum fluctuations at the teat end. Severe mastitis-causing vacuum fluctuations may occur at the teat end despite only minimal fluctuations at the pressure gauge on the vacuum line. (Courtesy of Dr. Steve Spencer, Pennsylvania State University)

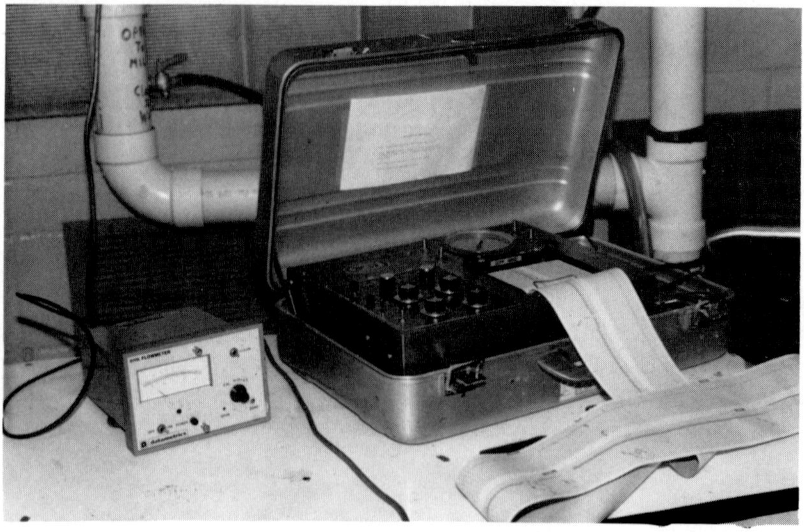

Maintaining the Milking System

You benefit from understanding how your milking system works. With this knowledge, you should be alert to anything unusual. Are the teat cups falling off? Are your cows comfortable being milked? How long does it take your system to recover if a cow kicks a unit off or it comes off when you are attaching units? You are responsible for maintaining your system for proper function. How long has it been since your system has had more than a casual inspection?

Use only the pump oil recommended by the manufacturer of your system. Oil must be changed regularly and added as needed. Keep belts and pulleys free of oil and grease. Maintain proper belt tension and keep spare belts on hand for quick replacement when wear has been excessive. Regularly clean your pulsators and have your serviceman check your pulsators twice a year.

Vacuum hoses and lines require regular maintenance also. Check air tubes and brush them weekly. Clean them as needed. Don't repair them with tape, and have replacements on hand when you spot a leak or bad hose. Stall cocks may be leaking. Point out these leaks to your serviceman. Keep valves and gaskets in good repair, not only to stop leaks but to keep your plate loop count from rising.

Check the vacuum controller and moisture trap weekly. Cleaning the filter on the vacuum controller is mandatory if you hope to keep it working efficiently. Have your line voltage checked by your serviceman or electrician. If voltage drops are characteristic of your system, perhaps you are overloaded or have a motor that needs replacement.

Inflations require regular maintenance. Keep an extra set on hand. Inflations last longer if you use 2 sets and alternate them a week at a time. Give the alternate set a rest after thorough washing. The set will regain some elasticity during the off week, prolonging life. Watch closely for signs of wear, such as bulges, cracks or holes and install new inflations when you see evidence of wear. Follow the manufacturer's recommendations on changing inflations and maintaining the short milk tube and pulsator tubes as well.

Stray Voltage

An astute operator spots problems associated with stray voltage early on. If you normally feed in the parlor, you may notice that your cows are not eating in the parlor. Milk letdown may be poor, and the cows may prance around and act up. Production gradually drops off and you may or may not have a high SCC.

Diagnosing this common problem may not be easy because of the number of potential places that stray voltage can originate. Consult your electrician and your serviceman for assistance.

If, during your routine equipment evaluation, you lose confidence in your local service people, don't just complain about them and give them a hard time. Consult with an outside expert. By doing this, several things will happen. Hopefully, you will regain confidence in your local service personnel because you found they are doing your work correctly. If they aren't, then you want to know what should be done differently and correct the problem. A nearby university often can refer you to experts for a second opinion. Your vet is also a good source of information.

A milker specialist can evaluate the system from 2 aspects. He should be able to measure the performance of the system, and evaluate the effects on udder health. Hopefully, you have maintained good relations with your local serviceman so that the specialist and serviceman can discuss the changes needed. The specialist may also reinforce the fact that the serviceman has done a good job.

Before you saw a limb off behind you, consult with your vet, equipment dealer, dairy extension specialist and dairy field man on the options you are considering. Beware of "shade-tree" engineers for recommendations on your equipment.

"I think you need an electrician here, rather than a veterinarian."

17

Milking for Maximum Yield and Optimum Udder Health

Few responsibilities on your dairy are more important than milking cows, yet it's too often taken for granted.

I'll Be Right Back

I am always reminded of Ned, a local dairyman who quit shortly after I started my practice. He couldn't make a living milking cows, so he drove a school bus to augment his income. Ned had a brindle cow that was his pride and joy, but he rarely mentioned her shortcoming: she was a hard milker. He hung a Surge bucket with a strap on her. This cow was so slow milking and his milker so worn that Ned drove his bus route while the milker hung on her! But she wasn't naturally slow milking. She had been conditioned to let her milk down slowly to eliminate any pain caused from vacuum on an empty udder.

Regardless of how Ned felt about milking, his milking routine was of critical importance to his dairy operation, and it showed. No matter how you cut it, Ned was harvesting his crop in a leaky bucket. Could this be the case on your farm?

Your Attitude is Important

Some dairymen find pleasure in the relative peace and quiet offered by the milking routine. They relate to their cows on an individual basis and are rewarded by seeing a cow's progress. Other dairymen regard milking as sheer drudgery. They find little pleasure in the monotonous routine of an assembly line of teats, udders, legs and dirty tails.

The way you milk affects both your yield today and in the future. While poor and inadequate milking procedures put some milk in the tank, you might as well do what you can to harvest more. A crop farmer wouldn't go into the field in the fall and pick only 2 out of every 3 corn rows.

Your challenge is to adopt milking procedures that maximize milk letdown, minimize the spread of mastitis-causing pathogens, and promote good teat and udder health.

Evaluate Your Milking Routine

Begin by taking a long, hard look at your milking routine. You may have been milking cows for 30 years and think you know all there is to know about milking. But the odds are excellent that you can make improvements. We all become creatures of habit, chained to techniques more beneficial to us than to the cows. You might want to ask your veterinarian, dairy extension specialist or dairy equipment specialist to help in this evaluation. Be open minded. If you discover you need to make major changes in procedure, make them one at a time. Some dairymen like to make up a timetable for these goals.

The Physiology of Milk Letdown

It's a fact of life that you can't take milk from a cow — you can only coax her into giving it to you. Between milkings, about 90-95% of a cow's milk is stored in the microscopic milk-secreting alveoli of her udder. When a calf sucks her or you milk her, milk is "squeezed" from her alveoli by the contractions of tiny muscle fibers around each alveolus. The action of these myoepithelial cells forces milk into the duct system, where it flows by gravity and pressure into the udder and teat cisterns. Without the squeeze, little of the milk stored in the alveoli could be obtained. It's the combination of pressure within the udder and vacuum in the milking system that makes it possible to remove a cow's milk.

Myoepithelial cells must be triggered to perform the squeeze. Here's how it all works: Stimulation by the calf or premilking preparation signals the pituitary gland to secrete a hormone called oxytocin. The oxytocin then moves out into the bloodstream, signalling the udder for milk letdown.

While cows of native, low-producing breeds require suckling by their calves or the presence of their calves to let down their milk for hand milking, cows of today's genetically refined, high-producing breeds let down their milk in response to a wide range of stimuli. Letdown is no longer just an "unconditioned" response to suckling, but also a "conditioned" response to many of the positive events and stimuli consistently associated with milking. These include grain feeding just before milking, washing and drying of the teats, forestripping, massage of the teats by the pulsating liners, and such sounds as the rattling of utensils, hum of the vacuum pump, snapping of pulsators and clanging of buckets. So-called "easy-milking" cows let down their milk solely in response to familiar sights, sounds and smells before they enter the parlor or before you prep them in their stalls.

Because oxytocin causes the myoepithelial cells to contract, pressure increases within the udder. A cow's milk is more "available" for removal while her blood oxytocin level is high. She'll milk out quickly and more completely. Though milking while oxytocin levels are high has benefits, taking advantage of high oxytocin levels isn't easy. The action of oxytocin is of relatively short duration: no more than 5-7 minutes. To take maximum advantage of oxytocin you must begin milking as soon as letdown occurs. This is sometimes impossible to do with cows standing outside the parlor, leaking milk in response to the sounds of pulsators snapping and stall gates opening and closing!

Unfortunately, some cows don't let down completely, even after the best teat and udder stimulation. If a cow is excited or frightened in the 30 minutes before milking, she may not release oxytocin when prepped. Excitement and fear cause the release of adrenaline, a hormone secreted by the adrenal glands. Adrenaline interrupts the letdown process by preventing the release of oxytocin and by interfering with its effects on the myoepithelial cells. Though they may not be hair-raising to us, loud noises, strange smells, stray voltage and foreign objects or animals can all trigger the release of adrenalin in cows and stop the letdown process. Unless you keep your cows calm before milking, milking will take longer and you'll limit the amount of milk the herd produces.

Maximum letdown is critical both for milk extraction and milk production. Because milk secretion is a continuous process, you need to remove milk from the udder frequently and completely to achieve high production. When a cow's udder is empty, her secretory cells produce milk as quickly as nutrients are provided to them by the blood. The manufacture of milk slows as the udder fills and pressure increases. If you remove milk periodically and udder pressure is relieved, milk secretion continues. If, on the other hand, you fail to remove milk, excessive pressure within the udder shuts down the secretion process. Secretion usually ceases when pressure in the udder reaches about 1.4 inches of mercury. Put simply, a cow can produce more milk than her udder can store. Your job is to remove milk often enough to keep your cows' secretory cells working for you.

Cows' udders vary in the amount of time they take to fill. The udders of cows in early lactation fill more quickly than those of cows in late lactation, and small heifers' udders fill more quickly than large cows' udders. Pressure within the udder begins to limit the rate of milk secretion 6-10 hours after the last complete milkout. The longer the interval between milkings beyond 10 hours, the more that production is slowed on a daily basis.

Milking twice a day does not significantly reduce production as long as the between-milking interval is roughly 12 hours. However, shorter intervals have a positive effect on production. Milking 3 times a day rather than twice a day increases production by 10-20%. Milking 4 times increases production by 25-30%. Remove as much milk as possible at each milking to maximize secretion between milkings. This is especially critical when the milking interval is 12 hours or more. Even with the best milking procedures and maximum milk letdown, it's never possible to remove *all* of a cow's milk from her udder. A certain percentage — usually 10-12% — is held in the alveoli unavailable to the milker. For example, even with good stimulation and normal oxytocin levels, a cow with 25 lb of milk will only give 18-23 lb.

The amount of residual milk left in a cow's udder depends on the extent of milk letdown, which in turn depends on stimulation and oxytocin release. Research has shown that the operator who provides enough stimulation during milking to promote maximum letdown leaves the least amount of residual milk in the udder. That is, he increases milk production and yield. In studies of twin cows, a cow that receives more premilking stimulation than her twin milks out more quickly, yields significantly more milk and fat, and has more persistent lactations. Evidence indicates that good, regular stimulation not only increases the amount of oxytocin secreted but also improves the persistency of a lactation by raising prolactin levels in the blood. Prolactin is the hormone that maintains the lactation.

You'll not be surprised to learn that procedures that fail to promote oxytocin release result in incomplete letdown and higher levels of residual milk. Studies show that consistently leaving more than 10% residual milk in the udder slows milk secretion, reduces milk and fat yields, and causes cows to dry off earlier.

Provide a Profitable Milking Environment

Though we ask cows to do a lot of "unnatural" things, such as eat prepared rations and live in stalls in crowded barns, by far the most unnatural behavior we expect of them is to give their milk to a machine twice a day in a noisy and strange environment. The smoother the routine for bringing cows and the milking machine together and the quieter the milking environment, the better the cows' letdown and the better their yield.

At milking time more than at other times, it pays to remember that cows are somewhat skittish animals. Summon up all your patience and understanding in dealing with them, because they produce best in a peaceful, predictable environment. To get the most from them, you should put *their* needs ahead of yours. Charm them into giving you their milk (Fig 1).

Add predictability to their lives by developing feeding and milking schedules and sticking to them. Keep the quality of their feed consistent. Above all, minimize sources of pain and excitement in their environment, especially before milking. Driving cows too vigorously, using barking dogs to herd them, allowing cows to crowd against electrically charged gates, and introducing new cows to the herd may all contribute to poor letdown and incomplete milkout.

Figure 1. Special soothing care, like music from a Stradivarius, will charm more milk from your cows. If you want her to give you what she has, you have to charm her into it!

Pick the Best Operator

Of course, a major factor in the milking environment is the person who milks. Studies have shown that cows produce more when catered to individually with good milking procedures. If you are lucky enough to be able to choose who will do the milking in your operation, select the person who is the best-trained and most competent milker. An operator who uses good procedures with poorly operating equipment will actually do a better job of milking than an operator with poor techniques and properly operating equipment. If your most competent milker is not your most enthusiastic one, see that the more enthusiastic milker receives the additional training he or she needs to do the job. Milkers who enjoy working with cows and your particular breed do a better job of milking.

Once you have decided who will do the milking, have that person do all of the milking. This will add consistency to the milking routine. It also enables the milker to become familiar with each cow's characteristics and milking habits. The operator also becomes sensitive to how the milking machine is functioning.

Evaluate Milking Procedure

When you're faced with a chore that must be done twice every day, there's a great temptation to race through it with only one goal in mind: to get it over with. Racing through some chores on a dairy farm may have few ill effects on the herd, but milking is the exception. The first step to a better milking routine and a heftier milk check is to slow down when you enter the parlor or barn to milk. The 15 minutes or so that a cow spends being prepped and milked each day are probably the most critical minutes of her lactation. If you prepare each cow with care, she'll milk out quickly and completely, her udder will be mastitis-free, and she'll have healthy teats and a low somatic cell count. Better yet, you'll have an annual yield you can brag about!

Treat a cow poorly and she's likely to pick up mastitis-causing bacteria and develop new infections. Bacteria then travel to uninfected quarters and the cow develops a high somatic cell count. She also fails to let all of her milk down, and can't meet her production potential. You don't necessarily have to spend a lot more time milking your cows; you just have to use your time wisely. Milking is the best method to prevent new mastitis infections and promote udder health while filling your bulk tank. I recommend the following procedures to achieve these goals.

Sanitize milk-handling equipment: To avoid contaminating your milk with bacteria that cause spoilage, and to avoid exposing your cows' teats to teat cups harboring mastitis organisms, sanitize all milk handling equipment about 10 minutes before you begin milking. Rinse the lines and equipment with clean tepid water containing an approved sanitizing agent, and allow them to drain and drip dry. Be sure to follow state and local regulations regarding the cleaning and sanitizing of milking equipment.

Milk cows in order: Because mastitis organisms (particularly *Staph aureus*, *Strep ag* and *Mycoplasma bovis*) can be spread from cow to cow by a milking machine, the likelihood of spreading mastitis from infected to uninfected cows can be reduced by milking infected cows last. Some experts recommend milking heifers (usually not infected) first, and infected cows last. Others suggest dividing the milking string into low-, medium- and high-somatic cell count groups, and milking from low to high. On some farms, dairymen milk infected cows in a separate "hospital string." And in large herds a second, smaller milking system with a separate receiver and bulk tank may be justified for milking treated cows.

These milking order recommendations make sense from a scientific standpoint, but not from a practical one. Milking cows in a particular order in a stall or stanchion barn is feasible, but following a particular order in a parlor is likely to pose major problems. In many herds, half of the cows are infected. Identification and segregation of these cows can be time-consuming. Extensive segregation of cows is also inconvenient if not impossible on your farm unless you're already grouping cows.

In most cases, it's safe to say that proper premilking washing and sanitizing of the teats and postmilking teat dipping reduce the need for milking cows in a particular order. You can also reduce the spread of pathogens from infected to uninfected cows by using an automatic

Figure 2. Newer milking machine designs include backflushing systems that cleanse the insides of the cluster with a sanitizing solution after each cow is milked. This controls the spread of mastitis-causing bacteria from cow to cow. (Courtesy of Germania Dairy Automation)

backflush system that rinses and sanitizes clusters between cows (Fig 2).

Wash, sanitize and dry teats: The first step in the actual milking routine is to wash, sanitize and dry the teats. I place emphasis on the word "dry" here because recent research shows that to prevent mastitis, you should only wet the teats and adjacent parts of the udder. Then you should get them bone dry before you attach the milker.

The Grade-A Pasteurized Milk Ordinance, used by inspectors to determine grade-A status on dairy farms, requires only that "the teats and udders of all milking cows shall be cleaned and treated with a sanitizing solution just prior to the time of milking and shall be relatively dry before milking." This requirement dates back to 1978 and is intended to protect the quality of milk that fills the bulk tanks of grade-A farms. It may reduce the amount of sediment and bacteria in grade-A milk, but it falls short of ensuring udder health.

Because bacteria in the environment make their way from outside the teats and udder to inside the teats during the milking process, it's important to know how to fight them. This group of bacteria includes the coliforms (*E coli*, *Klebsiella* and *Enterobacter*) and the non*agalactiae* streptococci. Unlike the "contagious" bacteria (*Strep ag*, *Staph aureus*, etc), which cause many subclinical (inapparent) mastitis infections, most infections caused by the environmental pathogens are at the clinical level. Unfortunately, the antibiotics formulated to treat both lactating and dry cows are effective primarily against the contagious pathogens. They are of little or no value in treating coliform infections and are only reasonably effective against the environmental streptococci. For this reason, the key to controlling infections by environmental pathogens lies in *prevention* rather than treatment.

This is where cleanliness is all-important. These organisms are best combatted by milking only clean, dry teats, and by maintaining a clean cow environment. Your primary goal during premilking preparation should be to eliminate as many bacteria from the teats as possible. The fewer bacteria on a cow's teats when you attach the unit, the less vulnerable she'll be to these kinds of environmental infections.

Maintaining cows in a clean environment between milkings not only reduces the risk of bacteria invading the teat canals while cows are feeding and resting, it also reduces on the amount of premilking preparation you must do. Washing mud and manure from teats and udders at every milking takes time. If your cows spend the day with their udders submerged in mud, you might better spend your teat-washing time cleaning the facilities. Scrape lots and lanes, change bedding, and build stalls that, by design, will keep your cows cleaner. Though clipping may not be required by your state health department, regularly clipping your cows' udders and flanks also keeps mud and manure from clinging to these surfaces.

The reason there's so much emphasis on drying udders and teats is that bacteria thrive in moisture. If a cow's udder isn't bone dry before you put on the milker, dirty water drains down the udder and collects around the teat and liner mouthpiece. This moisture is sucked into the liner and drawn down to the teat end where, if conditions are right, bacteria in the water invade the teat canal. Dry teats also provide a better grip for teat cup liners, so they're less prone to slip or fall off.

It's nearly impossible to thoroughly dry a wet udder during the average milking routine, so avoid wetting the whole udder. Wet the teats and an area no more than 2 inches above the base of each teat. Use only as much udder-washing solution as you need to get the teats clean. If you milk in a parlor and wash with a hose, direct the hose at only the teats or else use it to wet single-service towels. In a stall or stanchion barn, wash with single-service towels that have been immersed in a pail of clean udder-washing solution. Change the wash frequently during milking.

Single-service towels are more expensive and may be less convenient to use than udder sponges or cloths. But they're highly effective in removing bacteria and preventing the spread of bacteria from cow to cow. Udder sponges and cloths are virtually impossible to keep clean. Bacteria can survive in them even after you've soaked them in sanitizing solution for hours.

Quite a number of dairymen tell me that paper towels are a nuisance to use. So without charging a fee, I do a culture on their sponges or cloths. I don't have to say another word when I show them the results. The culture always reveals a number of different mastitis-causing organisms. When you consider the "costs" of mastitis, single-service towels seem like cheap and convenient insurance.

A number of udder washing solutions do a good job of killing bacteria on teats when you use them properly. To protect your cows' udders and teats from chemical injury and ensure that the teats are being sanitized, use only solutions intended for use as udder washes. Be sure to follow the directions on the label. Because most sanitizing solutions lose their ability to kill bacteria as they are exposed to dirt, it's important to periodically change the udder wash in your wash pail. Iodine sanitizing solutions are particularly useful because their color indicates their level of effectiveness. As long as an iodine udder wash is a light brown or tea color, it has enough strength to kill bacteria. When the tea color disappears, the wash is no longer able to sanitize and should be discarded.

For the sanitizing solution to have adequate time to do its job, hand wash each cow's teats and the udder surrounding the teats for at least 15-20 seconds. Your solution should be warm (about 110-120 F). The massage cleans and stimulates milk letdown. Then dry the teats and udder for at least 10 seconds. Drying with single-service towels not only removes any moisture but also physically removes bacteria from the teat skin. Don't be stingy. Be sure to use enough towels to dry all wet surfaces *completely*.

Cows also benefit from predipping teats. If you've kept your cows clean, this may be the only premilking prep they'll need.

Prestrip foremilk from each quarter: The earlier you detect mastitis, the easier it is to treat and the less likely it is to permanently damage delicate udder tissues. If your herd is on DHIA test, you can monitor the level of irritation in each of your cow's udders by watching their somatic cell count reports. By far the best method of early clinical mastitis detection is to prestrip each quarter on every cow at every milking.

After you wash and dry the teats, hand strip 3-4 streams from each quarter. If you milk in a tie stall or stanchion barn, strip the foremilk into a strip cup or onto a black plate. If you milk in a parlor, strip onto a clean portion of the cow platform or onto a dark tile set into the platform surface.

For those of you who are rookie milkers, here's how to strip just the foremilk. Use your thumb and forefinger to block the area of the anular fold (between the teat sinus and the udder) before you express milk through the teat canal. Check each milk sample for abnormalities. Be sure there is enough light where you milk to allow you to inspect the stripping carefully. If you detect abnormalities, separate the infected cow from the herd for further examination and treatment. Be sure you clean your hands and any contaminated equipment before you continue milking.

Removing the foremilk by hand has 2 other important benefits in addition to being the best means of detecting mastitis. First, hand stripping provides further stimulation of the teats to promote letdown. Ideally, you should prestrip immediately after washing and drying so the teats are continuously stimulated for a minute or more. Second, the foremilk you remove has a high concentration of bacteria and somatic cells, so it's better to discard it than to put it into the bulk tank.

Predip teats: Predipping teats, that is, dipping teats in a germicidal teat dip before attaching the milking unit (or spraying them with dip, using a specially designed nozzle), is a relatively new way to control mastitis. It was first tested on large, commercial dairy farms in California. It appears that predipping helps prevent new udder infections by reducing the bacteria on teats. This reduction in bacteria is less than the level achieved by washing teats with a sanitizing udder wash.

Scientists are studying predipping and several other measures that may fight the coliforms and other environmental pathogens cows encounter between milkings. Because control over contagious pathogens is improving and the use of confinement housing is increasing, there's now a greater threat posed by environmental bacteria. There's evidence that mastitis caused by environmental bacteria is increasing in most herds.

Research on predipping with an iodine-based dip indicates that this practice effectively reduces the number of bacteria found on teats *after* milking. Fewer bacteria are found on predipped teats than on teats not predipped. The higher the concentration of iodine in the dip, the fewer the number of bacteria found on the teats. Measurements of bacterial populations made after milking give a good indication of the ability of predipping to prevent new infections. Most new infections occur at the end of milking, when milk flow is low.

Studies also show that predipping with an iodine-based dip significantly reduces the rate of new infections by environmental bacteria. Where predipping has been compared with "no udder preparation" and "good udder preparation" (washing with single-service towels and an udder wash), the rate of new infections in predipped cows has been significantly lower than the rates resulting from either no preparation or good preparation. In at least one study, predipping after washing reduced the rate of new infections by more than 50% as compared with the rate associated with washing alone. Coliform infections were reduced by 54% and environmental streptococcal infections by 48.2%. This is what researchers found, but on most dairies, milkers don't do both.

Though predipping appears to have impressive potential as a weapon against environmental pathogens, it has one serious drawback: it can leave undesirable levels of residues in milk. Dipping teats in a 1% iodine-based dip both before and after milking can increase iodine residues in milk to well beyond the level tolerated by the FDA. When you predip instead of washing teats, sediment in milk can also be a problem.

Fortunately, you can control residue and sediment levels by washing teats before dipping, using dips with lower concentrations of iodine, and drying teats after predipping. Research on predipping residues has shown that drying teats thoroughly after dipping in a 0.1% iodine dip leaves less iodine in milk than dipping teats in the same solution after milking. Of course, you can eliminate residues by avoiding dips containing the kinds of compounds the FDA finds unacceptable in milk. At least one dip manufacturer (Diversey-Wyandot) has developed a dip for use both before and after milking with an active ingredient that occurs naturally in milk. Preliminary data indicate that the dip is highly effective in reducing new infections by environmental pathogens. Residues don't seem to be a problem.

Until researchers do more work on predipping, and the industry settles the controversy over residues, I can't give all herds a blanket recommendation on predip selection. However, if your herd has a problem with coliforms or environmental streptococci, you may want to try one. Consult your vet. If you elect to predip, wash your cows' teats before dipping and immerse at least the lower half of each teat in the dip. To prevent residue problems, use an approved dip containing active ingredients that are acceptable in milk. For example, 0.5% iodine effectively prevents new udder infections without causing objectionable residues. Leave the dip on the teats long enough for it to do its job (20-30 seconds), then dry the teat thoroughly with single-service towels.

If you elect to predip, tell your dairy fieldman that you're trying the new practice so he can be on the lookout for residues. Also, watch the rate of new udder infections in your herd. Predipping is a means of preventing mastitis, not curing it, so it'll have no effect on the udder infections your cows already have. The benefits of predipping may not be immediately apparent, but you should see a decline in the rate of new infections over several months.

Attach the milking unit: For years dairy scientists have recommended attaching teat cups roughly 60 seconds after the beginning of the washing, drying and prestripping routine. The idea was that it takes 45-60 seconds for oxytocin to reach the udder for maximum milk letdown. Some have argued that 1 minute of continuous stimulation (premilking preparation and massage) is necessary for full letdown, while others have said that 30 seconds of preparation followed by a 30-second delay before attachment will suffice.

The research that spawned the 1-minute-to-attachment recommendation was conducted in the early 1940s on Jerseys in New Zealand and Friesians in East Germany. When Jerseys were manually stimulated for 30-45 seconds, their annual yield increased by 25%. The yield of the Friesians increased by 12% under the same regimen.

Recent research indicates that the milk letdown response is now triggered more easily and that extensive manual premilking massage is not as necessary as it apparently once was. Now it's clear that you can take from 20 seconds to 3 minutes to get the unit on, depending on your milking system, without significantly reducing milk yield. In studies of high-producing Holsteins at various stages of lactation, differences in milk yields ranged from only 0-2% when the researchers compared no manual premilking stimulation to 30-60 seconds of stimulation. Though premilking stimulation has a relatively small effect on yield with some cows, it does increase the rate of milk flow and decrease the amount of machine-on time, especially during the last half of lactation. A decrease of 30-45 seconds in machine-on time results in less injury to a cow's teats.

Because there's so much variation in the type and extent of stimulation needed to trigger maximum letdown in modern cows, it's almost impossible to recommend any one stimulation and attachment routine for all cows. Cows of different breeds vary in the type and extent of stimulation they require, as do cows at different stages of lactation and different ages. The best rule to follow for premilking preparation and milker attachment is to treat your cows like the individuals they are.

For most cows, effective washing, drying and prestripping provide adequate stimulation to start letdown. There is little benefit in spending additional time massaging the teats or delaying attachment of the milker. The initial milk flow rate is sufficient for safe attachment of the milker. Stimulation begun during premilking preparation continues with the action of the pulsating liners. Cows that begin leaking milk before they are prepped should be washed, dried and prestripped quickly (though thoroughly), and milked as soon as possible. Others, such as low producers near the end of their lactation, may require more manual massage to let their milk down. The trick is to learn what each cow requires to fill her teats and then take the time to give her the stimulation she needs.

To take full advantage of the effect of oxytocin and to milk the cow out quickly and completely, it's important to attach the milker as soon as the teats are full of milk and *not before*. Though teat cup liners perpetuate the stimulation process by massaging the teats, most milking units are designed to be applied only after letdown is well under way. Attaching teat cups before teats are fully distended increases the incidence of liner slip (especially with small-teated cows). It also causes teat cups to crawl up and restrict the flow of milk from the gland cistern to the teat canal. Whatever milk is in the teat cistern and gland is removed quickly. Then full vacuum is applied to the tender walls of the teat canal and cistern, causing irritation and damage. Low milk flow and a "kicky," uncomfortable cow are clear signs that a unit has been attached too soon. Attaching it a little late, on the other hand, poses no major problem. You don't risk incomplete milkout unless you delay attaching the unit more than several minutes after maximum letdown.

Some milking units, called "stage" or "phase" milkers, are designed to be put on a cow before letdown occurs. These milkers massage a cow's teats with lower levels of vacuum until milk flow reaches a specified level. Then they deliver full vacuum for removing the milk. While these milkers may function properly on cows that haven't let their milk down, they can't usually accommodate cows that have.

If a cow has already let down her milk or you're delayed getting the milker on her, there's a sudden rush of milk into the unit before it can switch into the second, or milking, stage. The rush of milk may result in a drop in vacuum, causing the unit to fall off the cow. Because most cows let down their milk in response to necessary premilking preparation, "phase" or "stage" milkers have little practical usefulness with modern cows.

You can reduce the transfer of mastitis-causing bacteria from one cow to the next by handling and attaching the teat cups with care. When removing the cluster, prevent the teat cups from coming in contact with sources of bacteria, such as the parlor or barn floor. Try not to let the liner contact your hands, other milking equipment, or parts of the cow other than the washed teat. Attach the teat cups gently, keeping the liner mouthpieces well below the base of the udder and minimizing the amount of air that rushes into the cups. A sudden rush of air into the teat cups causes teat-end impacts and increases new mastitis infections. If you use suspension units or radial support arms, adjust the unit so the teat cups are aligned in the same direction as the teats and place a slight forward tension on the claw hoses. Positioning the unit properly facilitates milk flow and milkout while helping prevent teat-end impacts from liner slip and unit falloff. If you have plastic claws, you can check for milk flow from each teat. Occasionally, one test is inadvertantly doubled over and flow from that teat is shut off.

Monitor the milking system: Once you have attached the milker to a cow and the milking machine has begun its work, it's tempting to feel that you can forget about that particular cow for the next few minutes. After all, there are other cows to prep and milkers to remove, and a long line of cows waiting to be milked.

Though a cow may appear contented and safe while she's milking, she's actually at the mercy of the machine and in need of your attention. Every minute she's in contact with the machine, she's vulnerable to machine-induced infections caused by unstable vacuum and other machine malfunctions. Her vulnerability increases as she is milked out.

To reduce the chance of infection, maximize milk flow and minimize machine-on time, try to monitor cows while the machine is on. Be sure they're comfortable and that the milker is attached and working properly. Teat cups may fall or be kicked off and must be reattached. Sometimes they ride up on the teats and you'll need to bring them down. If the cow moves, you may need to reposition the entire unit to improve milk flow.

Milking time is the best time to detect problems with your milking system, as well. While you prep cows and attach and detach milkers, use your eyes and ears to evaluate the operation of the milking system. Listen to the pulsators at the start of each rest phase: listen for air hissing into the vents of the milk tubes or claw, and especially for liner squawks. Note cows that look uncomfortable and kick at the cluster. The teats may be bruised or blue. Look, too, for teat cups that ride up or fall off. It takes a sharp eye to spot patterns in milking machine malfunction that are limited to certain cows or certain clusters. If you remain alert to signs of trouble, you can detect and correct problems before they affect the health of your herd or the size of your milk check.

Remove the milking unit: The length of time required for a cow to milk out is affected by the cow's level of milk production, the degree of letdown, and her peak milk flow rate. When a cow is stimulated properly and letdown is complete, milk usually flows fast for the first 2 minutes. The cow can be milked out — with the help of her own release of oxytocin — in 4-6 minutes. In general, the longer it takes to milk a cow, the greater the amount of residual milk. Cows that milk very slowly are not likely to be high producers. If you can afford it, cull slow-milking cows. The odds are good that the calves of slow-milking cows will also be slow milkers, so you will be better off investing your time and money in better replacements.

If you milk with units that you remove manually, you must use judgment in deciding when to remove the milking unit from each cow. To obtain as much milk as possible and ensure the highest yield, continue milking a cow until milk flow nearly stops and the last quarter is milked out. To do this, you must have some visible means of determining milk flow, such as clear plastic short milk tubes or clear plastic claws. Checking flow by pinching the black short milk tubes is both unsatisfactory and unsafe. This provides only an inexact measure of flow and promotes teat-end irritation by interrupting vacuum flow.

Unless liners are slipping or squawking, it's better to leave all teat cups on while the last quarter milks out. Research has shown that 1-2 minutes of "overmilking" (that is, milking after milk flow has ceased) with a properly functioning milking machine does not damage teat or udder tissues. Removing individual teat cups with the vacuum on causes air to rush into the empty teat cups with bacteria-laden milk droplets to impact the opposite teats. Impacts occurring near the end of milkout are more likely to cause new infections because milk flow is too low to flush invading bacteria from the teat canal and sinus. If teat cups slip or squawk while the last quarter milks out, remove the unit and strip it out by hand.

Just as it's important to minimize the amount of air rushing into the teat cups when you attach a milking unit, it's also important to prevent air from being sucked in when you remove it. Always turn off the vacuum supply to a unit before you remove it. As air enters the air holes in the claw or short milk tubes, the vacuum level falls and the teat cups drop off into your hands. Never pull a cluster off a cow; only catch it as the unit falls off.

In the past, dairymen believed they needed to machine strip (massage the udder while pulling down on the claw near the end of milkout) to remove all milk at each milking. It was an attempt to get the highest rate of milk and fat secretion thoughout a lactation. Actually, it was developed as a means of restoring milk flow after liners crawled up teats and closed off the connection between the teat canal and gland sinus.

The development of narrow-bore liners and other improvements designed to minimize liner crawl have eliminated the need for machine stripping on all but "problem cows." Frequent machine stripping is also irritating to teat canal and sinus membranes, thus promoting mastitis. Liner slips are common during machine stripping, so try to keep it to a minimum. Limit stripping to only those quarters from which a reasonable amount of milk can be obtained, and make it rapid — no more than 15 seconds. If fewer than 4 quarters require stripping, prevent air from leaking into the empty teat cups of the unit while you strip those that do.

You may be able to avoid machine stripping altogether by discovering other ways to improve milk flow and milkout. Many factors, including teat cup design, cluster weight, operating vacuum, teat size and oxytocin level, influence how completely you can milk out a cow. If you review your milking procedures and evaluate your equipment, you're likely to find ways to improve milkout and improve udder health. Even your hardest milker may milk out more easily if she receives additional premilking stimulation at each milking.

When removing milking units manually, it's essential to have a prep attach-detach routine that enables you to get back to milking cows to remove units promptly. Though 1-2 minutes of overmilking with a properly functioning machine has no ill effects on the teats or udder, lengthy overmilking damages the internal linings of the teats and increases new infections. It also wastes time. Overmilking yields very little milk and indicates that the milker's routine is inefficient.

Dairy scientists have not established any particular amount of time after which cows are seriously affected by overmilking. It's clear, though, that overmilking is most detrimental with machines that don't function properly and on cows that already have infected quarters. Overmilking disrupts circulation of blood through the teat and reduces the effectiveness of the teat canal as a barrier against infection by causing erosions or pulling the inside of the canal out (pulling sphincter out). Signs of overmilking include pain during milking, red or bluish teats, and scarred or misshapen teats.

Teat-end impacts are more likely to cause new infections when milk flow is low because liner slip becomes more frequent as milk flow subsides. Overmilking results in cross-contamination between quarters. Though the percentage of cows that become infected as a result of lengthy overmilking is fairly low, overmilking has a major role in spreading infections from one quarter to another in cows that are already infected.

To prevent overmilking, develop a more efficient routine for prepping cows and attaching/detaching milking units. Cut back on the number of units you handle at one time. It may improve efficiency as well as shorten the amount of time you spend milking.

Milking units equipped with automatic takeoffs save time and enable a milker to handle more units. They also take the guesswork out of unit removal. Automatic takeoffs prevent overmilking by detaching the cluster when milk flow drops below a specified rate, usually 1/2 lb per minute. They also eliminate the need for machine stripping by milking until milk flow is very low.

However, automatic takeoffs have at least one shortcoming: they aren't sensitive to the status of individual quarters. After you remove a cluster, check all quarters to see if they are milked out, and hand strip those that are not.

Though automatic takeoffs offer several benefits, they can also be a prime source of new udder infections if you fail to adjust them properly. To prevent sudden vacuum changes and associated teat-end impacts, set the takeoffs so that vacuum to the unit is completely broken before removal. When you've got it right, teat cups begin to fall off before the retractor device is activated. Check your takeoffs periodically by disconnecting the cable or chain from the claw and observing the time of teat cup drop relative to the action of the ram.

Be aware that automatic takeoffs can also spread mastitis-causing bacteria by removing units in an inverted position and allowing the liner mouthpieces to drag over the parlor curb as the unit is retracted. To prevent contamination of liners, adjust takeoffs so that teat cups are suspended above all parlor surfaces.

Dip teats: After milking, a cow's teats and udder are particularly vulnerable to infection. The teat sphincter muscles that normally help prevent bacteria from invading the teat canal are temporarily weakened by the milk extraction process, so the ends of the teat remain open for some time after the teat cups are removed. Milk left on the teats attracts disease-carrying flies and provides ideal conditions for the growth of bacteria.

Years of research have demonstrated that dipping teats in an approved germicidal teat dip immediately after removing the teat cups is highly effective in reducing the rate of new mastitis infections. Dipping destroys pathogenic organisms on the teat and prevents growth of organisms within the teat canal. It also promotes healing of teat sores and prevents development of new sores, and leaves a protective germicidal residue on the teats between milkings. By reducing the rate of new infections, teat dipping reduces somatic cell counts and prevents destruction of valuable milk-secreting tissues by mastitis.

Germicidal dips contain different active ingredients, including alcohols, bronopol, chlorhexidine, iodine, quaternary ammonium compounds and linear dodecylbenzene sulfonate. Germicidal dips do not provide equal protection against all types of bacteria. While dipping reduces the rate of new udder infections involving *Staph aureus* and *Strep ag* (the 2 major mastitis-causing bacteria) by at least 50%, infections by environmental pathogens, such as *Strep uberis* and the coliforms, aren't reduced nearly as much.

Differences in the effectiveness of germicidal dips aren't caused by the inability of the dips to destroy certain species of bacteria but rather are due to differences in the origins of various pathogens and the ways they contact teats. *Strep ag* and *Staph aureus* are highly contagious organisms transmitted from infected to uninfected quarters primarily during milking. Germicidal dips applied after milking can kill most of these bacteria. But a cow's teats are typically exposed to the coliforms and other streptococci between milkings when the potency of germicidal dips is diminished.

A second type of teat dip called a "teat sealer" or "barrier-type" dip has been developed to provide better protection against this second group. Barrier-type dips are latex or acrylic latex-based products designed to coat the teats. They create a physical barrier against pathogens and may or may not contain a germicidal agent. Unlike some germicidal dips, they're gentle to the teat skin.

Barrier-type dips have one drawback: they're difficult to remove at milking time. And if you don't have the sealers cleaned off before you attach the teat cups, latex material may collect on weigh jars and receivers. Fortunately, these dips don't leave residues in milk, and vigorous cleaning can remove the latex from milking equipment. Some dairymen say the extra rubbing required to remove the dips from the teats also helps stimulate milk letdown.

Though teat dipping is one of the most effective means of preventing mastitis, it may take several months for it to substantially reduce the level of infection in a herd. The number of clinical and subclinical cases in a herd is a function of both the duration of existing infections and the rate of new infections. Teat dipping can control the rate of new infections, but it has no effect on the percentage of quarters already infected. The impact of teat dipping on the level of mastitis in a herd is much greater — and more obvious — when you use dipping in conjunction with treating dry cows and culling chronically infected ones. These practices reduce the number of existing infections.

To obtain maximum benefits from teat dipping, dip cows' teats immediately after you remove the teat cups at every milking. Use only an approved teat dip and cover at least the lower half of the teat with dip. Using highly acidic equipment sanitizers and udder washes as dips can cause severe teat-end sores that may predispose cows to mastitis. If possible, allow the dip to dry before you turn the cow out. If this is impractical, deliver fresh feed to the bunks before milking so your cows will stand and eat after they leave the parlor.

You can apply teat dips by either dipping individual teats in a container of dip or by spraying them with either a hand-held sprayer or a hose connected to a stationary pump and dip reservoir unit. Dipping is more common in the United States, but spraying is gaining favor. As a rule, dipping provides better protection than spraying because dipping usually results in better coverage. Hand-held sprayers tend to be cumbersome and deliver an uneven coating to the teats. If used properly, a hose and pump spraying system can be as effective as dipping while offering certain advantages over handling dip cups. The spray nozzles are easy to operate, you can deliver the solution at an optimal pressure, you can't spill the dip, and the reservoir requires fewer refills. Even more important, the solution you spray is fresh and uncontaminated by contact with the teats of other cows or the milking environment. Dairy equipment manufacturers are working on ways to automatically spray dips on the udders of cows as they move through the parlor exit lane or to apply dip through the teat cups at the end of milking.

For teat dips to be effective and safe to use, they must be stored and handled according to the manufacturer's directions. Store teat dip in a clean, dark, dry place that is not subjected to extreme temperatures. Freezing can cause the components of a dip to separate and the germicidal agents to become inactivated. Portions of the resulting mixture can be so acidic that they erode the ends of the teats and actually increase the cow's susceptibility to infection. Similarly, dips exposed to temperatures above 90 F may lose their effectiveness. Sodium hypochlorite and iodine-based dips are particularly vulnerable to high temperatures.

Like milk, teat dips have a "shelf life," and you'd be unwise to use them after their expiration dates. To avoid having to throw away expired dip, use containers of dip in the order in which you receive them. Don't buy too much dip at one time. Keep containers of dip closed to prevent evaporation, spoilage and contamination.

Teat dips are specially formulated solutions, so apply them at the recommended concentration. Never mix them with other products. Diluting a product below the level recommended does not save you money; it simply reduces or eliminates the dip's ability to kill bacteria. You can apply dips with any cup-like container, but it's best to use a container designed specifically for the job. The dip stays cleaner and you'll apply it more efficiently. Keep extra applicators on hand and remove sponges from applicators equipped with them. Sponges harbor bacteria that can inactivate the dip.

While milking, keep enough dip in the applicator to immerse at least the lower half of the teat. To maintain the dip's strength and minimize the spread of bacteria, use a separate dip applicator on infected cows, or milk and dip infected cows last. If the dip and applicator become grossly contaminated by dirt or manure or diluted by milk or water, discard the dip and wash the applicator thoroughly before refilling. After milking, make a routine of emptying and washing all dip cups. Though it may seem wasteful, throw away used dip. Pouring it back into the original container inactivates your fresh dip.

If you spray teats with dip, clean and rinse hose sprayer heads between milkings and check the mixing unit in the dip storage tank to be sure it's functioning properly. The heads and dip reservoirs of hand-held sprayers should also be washed and air-dried between milkings.

Though teat dips do considerably more good than harm, they sometimes have negative side effects. By their very nature, some are mildly irritating to the teat skin. Others can cause irritations by reacting unfavorably with premilking sanitizers and dips. To minimize teat irritations, avoid iodine-based dips that have a high phosphoric acid content. Also avoid hypochlorite dips containing more than 0.05% sodium hydroxide, and use sanitizers and teat dips that are chemically consistent. If you're not sure which chemicals can be used together safely, consult your vet, fieldman, or sanitizer and dip salesman.

Teat dips have one other drawback: they can promote chapping and freezing in very cold weather. While it's important to continue dipping during the winter months, take special precautions when temperatures fall below 10 F and conditions are windy. To minimize moisture on the teats, dip only the end of the teat and blot off any excess dip with a single-service towel. Using dips containing emollients, such as glycerin and lanolin, may help prevent chapping, but avoid those with high concentrations of emollients. This is because the germicidal activity of a dip may be reduced if emollient makes up more than 10-12% of the dip. Before turning cows out, be sure their teats are completely dry. Warming the teat dip slightly helps reduce drying time. To reduce the windchill factor, provide protective wind breaks around holding areas and keep fresh cows with swollen udders indoors. If temperatures and wind conditions become extreme, it may be advisable to stop dipping for a time.

Because there are so many teat dips on the market, picking the best dip for your herd can be a challenge. The 4 most important factors to consider in making your selection are FDA approval, proven effectiveness, active ingredients, and type of base. The FDA regulates dips only as to correct labeling. If a dip complies with FDA rules, the label on the container will state the name and percentage concentration of each active ingredient, directions for use, name and address of the manufacturer or distributor, production lot number, and expiration date.

Unfortunately, the FDA does not require proof of the effectiveness of a dip. The "burden of proof" is on the company that markets the dip. Specific teat dips have been studied in controlled research and found to be effective, but imitations are readily available. Even minor changes in dip formulations may reduce effectiveness, so beware of such products. To be sure you're getting your money's worth, request information from the manufacturer regarding compliance with FDA regulations and proof of effectiveness.

For advice on the active ingredients and type of base, consult your vet. Most of the active ingredients and bases on the market are safe and effective. The one major exception is some oil-based dips, which may actually promote mastitis.

Sanitize equipment and hands between cows: A sanitized milking unit is a short-lived thing. Once a "clean" unit is put to use it begins to accumulate bacteria both inside and out. It's quickly contaminated by milk from infected quarters, the skin of poorly washed teats, teat sores and bacteria-laden hands. And once contaminated, the milking unit may transfer mastitis-causing pathogens to every teat it touches.

The most obvious way to control the spread of bacteria by the milking unit is to prevent the unit from becoming heavily contaminated during milking. Though a milker inevitably acquires some bacteria when in use, you can limit the bacterial load by following the procedures on the next page.

- Sanitize the unit before milking.
- Milk infected cows last or with a separate cluster.
- Predip teats.
- Sanitize and dry teats thoroughly with single-service towels.
- Hand-strip the bacteria-rich foremilk from every teat.
- Dip teats after milking and sanitize your rubber gloves between cows.

Though time-consuming and sometimes expensive, sanitizing clusters between cows provides additional insurance against the spread of bacteria. Dipping the teat cups in a germicidal solution is the easiest and cheapest method of sanitizing, but it's also the least effective and most controversial approach. If you dip the teat cups properly (that is, 3 cups submerged at a time so that air can escape from the cluster through the fourth) for a long enough time in a frequently changed sanitizing solution, this practice may work. If nothing else, dipping in a clean solution may physically wash bacteria away. But when you don't dip the teat cups properly, they are actually contaminated with more bacteria by dipping.

Automatic backflush systems are highly effective in reducing the number of bacteria on liners. This equipment rinses, sanitizes with a germicidal agent, rinses and positive-air-dries clusters between cows. Research on the efficacy of backflushing has shown that it can reduce the number of new mastitis infections involving contagious pathogens, such as *Mycoplasma bovis*, *Staph* and *C bovis*. However, it has little effect, if any, on the rate of new infections caused by environmental pathogens, such as coliforms and *Strep* (except *Strep agalactiae*). This finding is not surprising because environmental pathogens are not commonly spread from cow to cow at milking.

The danger with both dipping and backflushing clusters is that both practices may give you a false sense of security. You still need your other mastitis prevention practices, such as sanitizing and drying teats and dipping after milking. Because of their cost, backflushing systems may only be justified with large herds in which a widespread outbreak of *Mycoplasma* or other type of mastitis could be devastating. Dairymen with small herds are probably better off concentrating their efforts on preventing contamination of their clusters.

Like teat cups, a milker's hands may become contaminated by mastitis-causing organisms very early in a milking. They become a major route for the transfer of organisms both between cows and from the environment to cows. Under practical conditions, it's nearly impossible to sanitize your hands because of the number of bacteria you expose them to while prestripping foremilk and handling milking equipment. I recommend that dairymen wear smooth rubber gloves and dip them in a pail of sanitizing solution or spray them between cows. While milking, it's important to minimize contact with contaminated objects, such as strip plates and barn equipment.

Clean equipment after milking: Give your milking equipment a thorough washup after each milking. It's yet another way to control mastitis, extends the life of your inflations, and helps you produce high-quality milk. Do this manually with brushes or automatically. With the latter method, you need adequate velocity and water temperatures to replace brushing.

Regardless of the washing method you use, your goal should be to produce clean surfaces that are free of bacteria and sediment. After washing, inspect your equipment to see if all surfaces are completely clean, especially crevices in which milk solids might collect. Mastitis-causing organisms thrive in milk residues left on your equipment and invade your cows' teats during subsequent milkings. The time you invest in checking over your equipment is a small price to pay for preventing a machine-induced outbreak of mastitis.

With clean-in-place equipment, prerinse the entire pipeline system with 100- to 110-F water until the water comes out clear. I don't recommend recirculating the rinse. Next, the equipment should circulate water containing a low-foaming chlorinated detergent at a temperature of 140-160 F. For the wash solution to be effective it must stay hot. Unfortunately, soils can be redeposited at temperatures less than 120 F. It's important to keep the temperature of the wash on the return side as close to 140 F as possible. If you have 3-inch pipelines, you must usually inject air to create enough turbulence to wash the upper halves of the lines.

After washing, the wash solution drains out. Then your equipment is rinsed first with clean lukewarm water and then water containing a liquid acid, such as phosphoric acid. The acidic rinse eliminates any accumulation of milkstone and improves rubber life. Make certain you hand wash and rinse all equipment that's not part of the clean-in-place system.

With clean-in-place equipment you must be sure you have a hot water heater capable of supplying the large quantities of hot water the system needs to do an effective job of cleaning. If you have expanded your milking system since you bought your heater, you may need a larger heater. Some states have specific size requirements for water heaters used with different types of cleaning systems. Many new household heaters are designed to reach a maximum temperature of only 140 F, so they're unsuitable for use with your milking equipment. Commercial heaters have a greater capacity as well as higher maximum temperatures.

If you have equipment designed to be cleaned manually, disassemble all parts to be cleaned. Rinse all parts in 100- to 110-F water and then soak them in hot water containing an alkaline detergent. Brush all parts to remove soil before placing them in a second sink to rinse with clean warm water to remove mineral deposits. Often a final acidic rinse may be necessary. After rinsing, store the pieces so they drain and drip dry.

Be sure to follow the manufacturer's instructions regarding detergent concentrations and washing temperatures and times. Also keep up on local health department regulations pertaining to milking equipment. If you have questions about cleaning your equipment, consult a dairy sanitarian or representative of your milking machine company.

Cleanup time is the best time to check your liners for signs of wear. Liners with cracks of any size are difficult to clean and sanitize; replace them. As liners are used, they also lose some of their massaging action. In most cases, try to replace stretch liners after 500-600 milkings, and molded liners after 1000-1200 milkings. Follow the manufacturer's recommendations if they vary from these guidelines.

18

Nutrition and Milk Production: Protein, Energy, Fiber and Buffers

The most important thing that I can instill in a dairyman is the importance of feeding dairy cows for maximum intake. This is commonly referred to as dry-matter intake. The more feed a cow eats, the more milk you're going to get. Dry-matter intake is directly influenced by the types of feed you're using.

If you're feeding high-moisture feeds, feed intake is limited by the feed's high water content. This effect is not very great until the water content of a cow's total feed intake is above 50%. Beyond that level, there is a direct reduction of the amount a cow can eat. Consequently, milk production is reduced when large amounts of high-moisture feed are fed.

Pregnant cows eat about 20-25 lb of dry matter every day before freshening. Upon freshening, they are thrust into the production routine and pressed to milk at full potential. When a cow is producing at full potential, she may eat 40-50 lb of dry matter every day. A change from 20-25 lb of feed a day to 40-50 lb is a huge increase in a short time.

During that time of transition, everything must be done to encourage her to eat a lot of highly palatable, high-density feeds (low in water and with many essential nutrients for their weight). Newly fresh cows benefit from dry hay and a proper balance of fiber and nutrients. Remember, just because she is willing to hog down lots of concentrate for you doesn't mean that she is getting enough forage outside on the bunk. She must consume about a 50:50 ratio of forage to concentrates if she is to reach peak production without a digestive upset. Sometimes a cow will not eat well at the bunk and yet eat all of the concentrate you put before her. If she does this, she will likely go into a tail spin and crash with a digestive disorder.

High-moisture feeds, which are of lower nutrient density, are not as beneficial to a newly fresh cow because she has low feed intake. To make the newly fresh cow produce at her optimum, get as many nutrients into her as possible, in the proper balance.

Chapter 18

Components of Feed

Protein

Protein sources are the first nutritional topic most of my clients want to discuss. To understand a cow's protein needs, it's helpful to understand the cow's digestive system.

The cow is a unique animal, with several compartments to her stomach. She can break feed down by fermentation and extract their nutrients (Fig 1). Or, she can pass the feed through the fermentation vat for direct digestion by the abomasum (true stomach) and small intestine. By virtue of having 2 systems for digestion, cows are very efficient in using food sources that other animals can't.

The fermentation vat is called the rumen. In the rumen are microorganisms that break down soluble protein into ammonia, and cellulose into carbohydrates for energy. The ammonia then is recaptured by the microorganisms, with the released energy reassembled into amino acids that the cow can use.

Because of the rumen microorganisms, cows can use nonprotein nitrogen sources. Nonprotein nitrogen is any chemical nitrogen source that is soluble. For example, some feed companies use urea to supply inorganic nitrogen for extra amino acid formation in the cow's rumen. The cow breaks down this chemical nitrogen into ammonia, and makes amino acids from it.

Figure 1. Ruminants have 4 compartments in their stomach. The rumen is a giant fermentation vat. The second compartment, the reticulum, works like a stone trap on a combine to prevent damage from metal swallowed in the feed. Third is the omasum, or "manyplies," which absorbs water from the feed. The fourth compartment, the abomasum, is a true stomach. It has hydrochloric acid for digestion and, under stress, can develop ulcers.

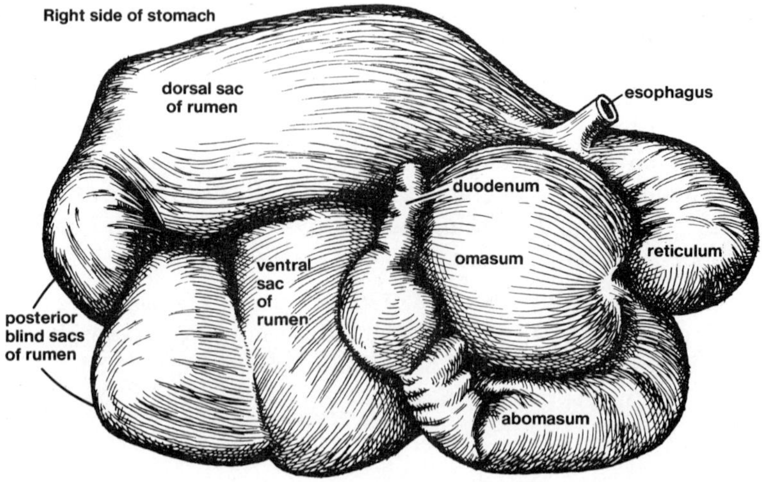

The cow has definite limits on the amount of nitrogen that her rumen can use. Some of the excess ammonia is passed off by belching, but most escapes through the rumen wall and is detoxified in the liver. If too much ammonia is released in the rumen, the ammonia that escapes through the rumen wall into the bloodstream may exceed the level that the liver can detoxify. If a cow accidentally eats extremely high levels of nitrogen, such as pure undiluted urea, the extreme amounts of ammonia released through the rumen wall will poison her. Without the proper antidote, death is very rapid.

How do you know when a cow is reaching her maximum ammonia level for soluble protein usage in the rumen? In large-breed cows milking less than 55 lb a day, milk production efficiently uses almost all of the ammonia released in the rumen. Above 55 lb of milk production, these cows may eat more soluble protein than the rumen microorganisms can break down. Above 55 lb of production, feeding extra soluble protein will do little or no good. In small breeds, maximum soluble protein utilization occurs at the 35-lb level.

There are a number of sources of soluble rumen protein. The only 100% soluble rumen protein source is nonprotein nitrogen, such as urea. It is totally soluble in the rumen and can be almost completely broken down and turned into amino acids. Other natural protein sources are highly soluble in the rumen, but none is 100% soluble like urea.

An example of a highly soluble protein source is soybean oil meal. Soybean oil meal, depending on the source, is usually about 80% soluble in the rumen. This means that 80% of its protein can be broken down into ammonia and most of it recaptured in amino acids by the rumen microorganisms. The other 20% goes through the rumen and on to the small intestine. Of the 20% that bypasses the rumen, most is digestible in the small intestine. A small percentage of it is totally undigestible and passes through the cow as manure. These solubility percentages depend upon the soybean crushing process used. Another example of highly rumen-soluble feed is barley.

The protein that escapes rumen fermentation is called bypass protein or insoluble protein. This protein is not broken down in the rumen. Rather, it is broken down in the small intestine and its amino acids are directly absorbed. If this protein has been bound in lignin, which is an undigestible form of carbohydrate, most of it cannot be used by the small intestine and it passes out in the manure. An example of lignin-bound protein is that found in soybean straw.

Soluble protein can be converted to bypass protein by heating or chemical treatment. Heating or chemical treatment changes the protein so it is not soluble for rumen action. The treated protein passes to the small intestine, where it is digested and directly absorbed. If the protein has been treated at temperatures above 300 F, it may not even be digestible by the small intestine, and is passed on in the manure. It is important to remember that high temperatures can damage protein. Protein may become undigestible if you have a silo that is leaking and the silage gets hot. Excessive heat damages the protein and makes it unavailable to your cows.

Why should you be concerned about feeding bypass protein? Because you could be missing a number of benefits. The benefits of using soluble and bypass proteins in your herd are increased production and increased profits. By balancing ration for protein and then fine-tuning with bypass protein, you may be able to increase daily milk production by 10-15% per cow. Often these bypass protein sources are less expensive than soluble protein sources. They can reduce out-of-pocket feed costs in addition to raising milk production. You also lower costs for each 100 lb of milk produced because of increased production.

As feed intake increases with milk production, more protein becomes available for rumen action. When a large-breed cow produces more than 55 lb of milk daily (over 35 lb for small breeds), soluble protein intake may be more than the rumen microorganisms can use. At this point, the cow can get a boost by using additional protein that bypasses the rumen and moves on to the small intestine. Additional protein fed for higher milk production must have lower rumen solubility so she can pass it through the rumen and use it in the small intestine.

Think of a cow in terms of a tractor with a turbocharger. At certain levels of performance, the tractor runs on conventional fuel and air for energy. At high performance levels, this tractor cannot meet those horsepower requirements with just diesel fuel and air. At this point, the turbocharger kicks in to pump extra preheated air into the system to help improve combustion.

The cow's turbocharger is her small intestine. When she is not milking heavily, her protein requirements are met adequately by rumen microorganisms. When she is milking above 55 lb, the bypass protein system is needed to supply the extra protein required for higher levels of milk production.

The usual sources of bypass protein are by-products feeds, such as distillers' or brewers' grains. Other feeds have been changed to bypass proteins by special treatment, such as with heat or formaldehyde. Some feeds are naturally high in bypass protein by virtue of the time of year of harvesting. For instance, early-cut haylage has much more soluble protein than late-cut, maturing haylage.

There are some precautionary rules in using a balance between soluble and bypass proteins. You should not feed *all* bypass protein sources, such as distillers' and brewers' grain. For up to 55 lb of daily production, rumen microorganisms use the protein naturally. If you do not feed soluble protein to meet that level, the rumen microorganisms do not thrive for optimum digestion of cellulose and maximum carbohydrate utilization. If you're short on soluble protein, the cow cannot properly digest her feed. High levels of bypass proteins should only be fed when production is higher than 55 lb (above 35 lb in small breeds), and soluble protein is already being fed to maintain this production level.

Bypass feeds sometimes have special storage requirements. Often the best buy is when the feed company delivers bypass feeds in semi-load lots. This feed may get stale or moldy if it is not fed in a reasonable amount of time, so you may not benefit from buying in large quantities.

The protein requirements in newly fresh cows are much higher than in cows in later lactation. Many newly fresh cows benefit from rations containing protein in levels as high as 19-22%. A 19-22% protein recommendation must be balanced cautiously in terms of soluble and bypass protein so that soluble protein levels are met but are not excessive. Requirements for cows in mid-lactation often can be met with a 17% protein ration, with 60% soluble protein and 40% bypass protein. Dry cows don't need supplemental bypass protein. The main thrust of protein feeding in dry cows is to provide adequate soluble protein or nonprotein nitrogen for the dry cow to keep her rumen microorganisms thriving.

To determine if you are feeding excessive soluble protein, your vet can do a blood test on a few of your high producers. A blood urea nitrogen test can determine if there are excessive urea levels in the bloodstream. You can then adjust your protein levels to get optimum utilization of rumen-soluble protein. High blood urea and ammonia levels also can interfere with fertility. There have been reports of conception rates dropping by half when cows ate excessive soluble protein. You can unknowingly begin feeding excessive soluble protein by changing forages and not making an appropriate adjustment in the protein source.

Energy

If you shortchange a cow on energy, the rumen microorganisms do not adequately recapture ammonia, even if rumen protein and bypass protein are balanced. Therefore, it is critical that dietary energy requirements be met. Energy in feed becomes available with the microbial action of the rumen and digestion in the lower intestinal tract. Cows can derive energy from such sources as cellulose, but simple-stomached animals cannot.

During digestion of cellulose, the rumen microorganisms convert the energy in cellulose to volatile fatty acids, including propionic acid, acetic acid and butyric acid. These volatile fatty acids are then used as energy sources. The amount of butterfat a cow produces is directly affected by the proportions of these volatile fatty acids. Top butterfat production occurs when acetic acid is at optimum levels. Propionic acid is the other major energy source for the cow. Much of a highly digestible feed, such as high-moisture corn, is converted to propionic acid. When propionic acid levels increase in relation to acetic acid, the butterfat test drops.

A cow needs a certain amount of energy for maintenance and replacement of depleted tissues and an additional amount for lactation. When calculating the dietary energy levels a cow needs, it is important to include both maintenance and lactation requirements (Fig 2).

One of the newer principles of feeding dairy cows is a term called regulated carbohydrates. Carbohydrates are the energy source for a cow. The concept of regulated carbohydrates refers to regulating use of carbohydrates in proportion to use of proteins. This is done to have optimum available energy along with the optimum available protein. Remember, protein use drops if there isn't adequate energy available to the rumen microorganisms. Regulating carbohydrates is a method of balancing energy availability to protein availability in highly soluble diets, such as haylage. Balancing carbohydrate solubility to protein

Figure 2. To breed efficiently, a high-producing cow must be in good flesh or gaining weight. A peak-producing cow is in a negative energy balance and loses weight.

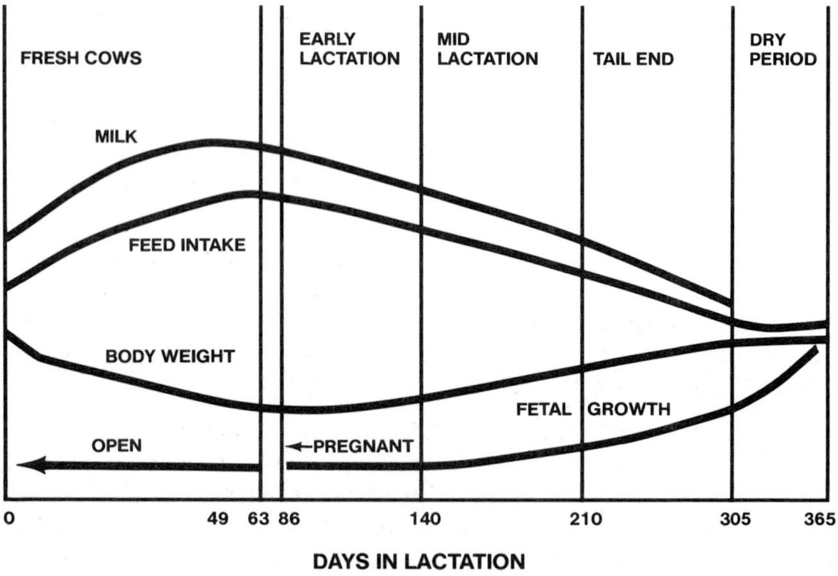

solubility sometimes can yield another 5-10 lb of milk production. If you are interested in using this principle of carbohydrate feeding, talk with your local feed company representatives or nutritionist.

A second method of enhancing energy is by using a group of compounds called isoacids. When fed in the correct ratio, isoacids enhance the rumen microorganisms' ability to digest cellulose and recapture ammonia. Consequently, protein utilization is improved. Isoacids also stimulate the cow to produce some additional bovine growth hormone. The results are a boost in production.

In principle, feeding isoacids appears very beneficial. It may be beneficial in some herds, but in other herds it has had no benefits. Consequently, isoacids have a controversial history. Research has not determined why they work well in some herds and not in others. One company manufactured them, but the product is no longer available.

Cows lose weight if they are producing milk at higher production levels than feed intake can support. In these instances, it is beneficial to feed a high-density energy source with the regular feeds to improve the energy supply.

Fat is a good energy source that contains 2 1/4 times the amount of available energy as compared with carbohydrates. It improves production by supplying extra energy without increasing dry-matter intake.

There are some disadvantages to feeding fat. The ration must be watched closely. Fat binds some dietary calcium and magnesium in the rumen, so additional calcium and magnesium may be needed. Butterfat sometimes drops. Fat also may make the feed rancid in hot weather.

Whether we are in favor of bovine somatotropin (BST) or not, it has changed the way we feed dairy cows. Cows given BST have higher dietary protein and energy requirements. When researchers learned how best to feed cows given BST, they also learned much about how to feed cows not given BST. In fact, cows fed according to this new information had about two-thirds the positive response of cows given BST.

Researchers have discovered that a dairy cow can use up to about 6% of dry-matter intake as fat. Most rations have about 2.5% naturally occurring fatty acids in forage and concentrate, leaving about 3.5% that can be provided as supplemental fat. This supplemental fat should be supplied as a combination of animal and vegetable fats.

A new fat source called "protected" fat eliminates most of the problem with rancidity. Also, it is unaffected by the rumen because it passes into the abomasum, where the cow can use it directly. The main disadvantage is that it takes a milk increase of 3 lb per day per cow to pay for its cost. In average herds, it won't pay unless the herd is having problems with thin cows and getting cows bred back. In these situations, it can be a real asset.

In your dairy cow ration, feed supplemental fat up to 6%, then use protected fats, such as Megalac, to provide additional energy for cows that require dietary fat levels above 6%.

Fiber

Fiber also plays an integral role with energy and protein metabolism. Cows eat large quantities of feed, regurgitate the feed later and chew their cud. Cud chewing creates copious amounts of saliva, which maintains the correct rumen environment for fermentation. When dietary fiber levels are low, the cow does not chew as much and thus secretes less saliva. Low-fiber, high-concentrate rations also suppress butterfat production.

Dietary fiber levels are measured with various analytical tests in a laboratory. The old standard measurement was crude fiber (CF). A cow should have 17% crude fiber in her diet. A newer measure for fiber is acid detergent fiber (ADF). Acid detergent fiber levels in the diet should be 21-22%.

A third measure of fiber is neutral detergent fiber (NDF). Neutral detergent fiber is a good method of comparing feed that is more like concentrate than roughage, such as succulent alfalfa. While there is much research going on with NDF, the results vary among the laboratories running the tests. Until some of these discrepancies are resolved, ADF is probably a more dependable measurement of fiber.

Fiber levels should be much higher in the diet of dry cows than in the diet of lactating cows. The principles of feeding fiber to dry cows can be best summarized this way: A lactating cow may eat 40-50 lb of dry matter per day. As she gets closer to calving, she is turned dry and fed a predominantly roughage diet. When she is dry, feed intake drops back to 20-25 lb of dry matter. This dry cow maintains herself on half the amount of feed needed for lactation. The problem is that if you provide her with high-energy feeds, her digestive tract does very little work to get the nutrients she needs in that 20 lb she eats every day. When she freshens, her digestive tract is not prepared to digest the large amount of feed offered to support milk production. Often the result is a displaced abomasum.

If a dry cow is to maintain good digestive tract function during the dry period, a high-fiber diet is necessary. Such feeds as longstem hay are not easily digestible, so they keep the digestive tract in shape.

Dry cows may require some extra energy and protein with the hay to meet nutritional requirements. It is no problem to feed extra protein and energy in the form of either grain or corn silage to meet these needs. The important point for you to remember is that dry cows should be fed a high-fiber diet. Such feeds as longstem dry hay, corn stover and oat straw are excellent feeds for dry cows.

Buffers

Rumen buffers help maintain the correct environment in the rumen and cecum for optimum fermentation and volatile fatty acid production. A commonly used buffer is sodium bicarbonate (baking soda). It can be fed up to 0.75% of the grain ration or 0.4% of the total diet. Magnesium oxide is another buffer you can use. It can be fed up to 0.5% of the grain ration. It is commonly used with sodium bicarbonate to get optimum buffering for top butterfat production.

A third buffer is sodium bentonite. It is commonly fed at 0.25-0.5 lb per cow per day. It is occasionally fed free choice, though this may cause some health problems. Sodium bentonite is a clay product mined from lake beds and has some buffering properties throughout the digestive tract.

Another dietary concern involves fermentation of starches in the cecum. This occurs with high-starch diets, such as high-moisture corn or high-energy pellets containing low fiber.

Cecal fermentation can be detected by analyzing the manure pH. Your vet can arrange for this to be done. Buffers, such as ground limestone, should be fed to correct the problem. Calcium carbonate, commonly called ground limestone, is an effective buffer on the lower digestive tract. Another option is to change the type of carbohydrates being fed so there is less starch in the diet. Specifically, decrease the amount of high-moisture corn or pellets in the diet and raise fiber levels.

You can easily spot a need for ground limestone if your cows are being fed a corn concentrate diet, as there usually is a lot of corn in the manure. *If you see excessive corn kernels coming through your cows, you can correct this problem by feeding extra calcium carbonate.*

Keeping 1% of the grain ration as ground limestone corrects the excessive corn in the manure of these cows. You can go as high as 2% of the grain to correct excessive levels of corn coming through in the manure. This also will correct the low manure pH.

It is impossible to describe the effects of improper buffering without including the effects of low-fiber, high-energy diets with it. Regardless of which one you wish to talk about, the signs and treatment are interchangeable. The results of low-fiber, high-energy diets with improper buffering are rumen acidosis and laminitis. Rumen acidosis suppresses the appetite and depresses butterfat test, resulting in laminitis. Laminitis is characterized by splay feet or long toes and deformed hooves.

The ultimate result of rumen acidosis can be a displaced abomasum or a crippled cow. This is why diets must be correctly balanced for energy, fiber and buffers. These are the circumstances in which a total-mix ration can be invaluable (see Chapter 20).

19

Nutrition and Milk Production: Minerals and Vitamins

Cattle require proper minerals and vitamins for production and reproduction. This is especially true of dairy cows. They need adequate levels of vitamins and minerals to reach optimum milk production and to calve every year. To meet these nutritional requirements, try to provide separate rations for dry cows, newly fresh cows, and high and low producers. Each group has different requirements, and separate rations make the job of feeding proper vitamin and mineral levels easier.

Mixing Minerals

Be aware that when you mix various vitamins and minerals together, certain elements may deteriorate. This is especially true of vitamins A and E. Some mineral manufacturers double the amount of vitamin E in their mineral packages to ensure adequate levels of vitamin E. Experts estimate that 50% of vitamin E is destroyed in feed processing. Mold in feed also inactivates many trace minerals, such as selenium.

Most roughages and concentrates contain some minerals, so you should balance your mineral program with this in mind. It's a tricky business because there is tremendous variation in availability of minerals to dairy cows.

Generally, minerals in the oxide form, such as copper oxide, aren't available for use by the cow after feed is consumed. Zinc oxide is an exception. Mineral sulfate and carbonates are more available. Many mineral anions interact also with each other and may bind into an unavailable form.

With dairy cattle, the only mineral that feeds well free choice is salt. Though thousands of dairymen feed the other minerals this way, I think it's a poor method of supplementation.

While it has been shown that cattle deprived of phosphorus crave bone meal to meet their phosphorus requirement, in at least one study phosphorus-deficient cattle on pasture ignored calcium and phosphorus supplements provided for them. When researchers changed the basic pH (acid-base composition) of these mineral supplements to acidic, the cows overindulged on them! Cows just cannot balance their diet properly for calcium, phosphorus and trace minerals on a free-choice feeding system.

Tractor power takeoff (PTO) grinders aren't suitable for mixing vitamins and minerals used in small quantities. All vitamins and minerals should be premixed with the protein source or other major additives before putting them into a PTO grinder. A horizontal mixer is the best tool for mixing small mineral ingredients effectively. Many of my clients use an old cement mixer to mix their micro-ingredients with the protein base mix before adding the complete mix. I strongly encourage you to carefully manage your mineral program this way. If you cannot premix your micro-ingredients in a horizontal mixer, then purchase manufactured mineral mixes that contain optimum levels for your cows.

The Importance of Water

Water is a very important nutrient for your cows! It is necessary for critical body functions, including cooling, nutrient absorption and excretion of metabolic wastes. Sufficient, clean water is critical for optimum milk production. An adult cow drinks 8-60 gallons per day. Cows in heavy production have an especially large requirement for water (see also Chapter 20). Your cows' water should be free of pollutants because there can be interaction between the cows' nutrients and water pollutants, other chemicals (such as nitrates), and coliforms. These can seriously affect a cow's ability to produce milk and to reproduce.

A Water Wager

Garth is one of my favorite clients. He milks 75 cows, and when he's not doing alot of custom harvesting, he takes good care of his cows. He told me he was concerned that his cows weren't producing like they should.

I had been working with Garth for a couple of years, had helped him lower his SCC and had tested his ration. I knew that we would have to look elsewhere for a solution to his low production problem. After considering several possibilities, I suddenly realized that his cows had only one automatic waterer. I told Garth that his cows couldn't get enough water. He didn't believe me. He quickly pointed out that the cows weren't congregated around the waterer. He also pointed out that it was January. If it was summer he might accept my suggestion.

The water out of the automatic waterer was warm, and I agreed that it didn't appear to be a problem. Nevertheless, I was confident that they weren't getting enough water. I proposed that Garth take the extra water tank he used in the summer for heifers on pasture and make it available to the cows. I also suggested that if I were wrong, I would buy Garth a steak dinner. I told him it was only fair that if I was right, he had to buy me one. Garth pounced on that offer like a cat on a mouse. During my bragging and running commentary, I casually mentioned that I would pick the restaurant. I went home figuring that we might as well go for the fanciest restaurant in town.

A week later, I was at Garth's and he asked me where I wanted to go for steak. His cows had come up 300 lb in milk on the first pickup after filling the water tank. I just grinned because I knew that cows nearly always prefer a tank over an automatic waterer.

Water doesn't contain a consistent level of the various minerals. Some water contains excessive fluoride, nitrates and nitrites. Saline, or salt, in the water may supply too much sodium for the cow. Sulfates can interact with molybdenum and create a complex that ties up copper. So a high level of sulfates in your source may cause your cows to have a copper deficiency (Fig 1).

Minerals

Calcium (Ca)

Background: Calcium is the most abundant mineral in a cow's body. Nearly all of it is stored in bones and teeth. About 1% is widely distributed in body fluids.

Function: Calcium is necessary for a cow's bone formation and maintenance. In body fluids it is a catalyst for nerve function, muscle contraction and relaxation (especially heart). Activates some enzymes. Part of mechanism to clot blood. Dry cows require limited amounts of calcium to maintain muscle tone.

Interactions: Ability to use calcium depends on vitamin D supply in body. Important relationship with phosphorus.

Sources: Readily available. Found in alfalfa, other legume forages. Supplements: carbonate, calcium sulfate (gypsum) and dicalcium phosphate.

Requirements: Dry cows: to maintain muscle tone (large breeds), no more than 100 g/day (small breeds no more than 60 g/day). Lactating cows (for example, 1300 lb) -- NRC established levels of 39 g (maintenance) plus 3.21 g/kg milk (lactation); I think values should be 25% higher – NRC has only established minimum values.

Signs of Deficiency: In lactating cows signs of deficiency include depraved appetite, low milk output. Vet may notice poor uterine tone. Cows slow to recover after calving. In young calves severely limited on calcium, swollen joints, leg deformities, bones susceptible to fractures.

Signs of Excess: Excess calcium may bind zinc in the diet (see zinc deficiency). In dry cows, milk fever is common at freshening (see also Chapter 7).

Signs of Wide Ca:P Ratio: If greater that 2:1 and phosphorus levels are deficient, a cow may have a depraved appetite, chew wood or eat dirt. May come into heat every 10-12 days or may not show heats, and drinks other cows' urine as it is voided.

Causes of Deficiency: Feeding a ration of mixed hay and corn silage without calcium supplementation. Fat added to the ration binding Ca. Must raise Ca when feeding fat.

Figure 1. This chart shows the interrelationships of minerals in your cows' diet. For instance, if a cow is eating a low-phosphorus diet, many trace minerals are adversely affected.

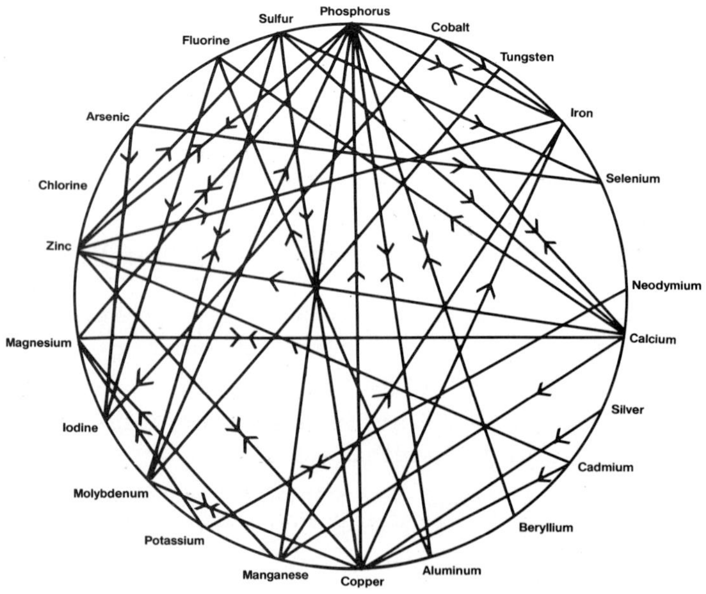

Causes of Excess: Failure to lower dietary calcium supplementation after a grass hay ration is replaced by alfalfa.

Causes of Wide Ca:P Ratio: When alfalfa (high in calcium) makes up major portion of ration and phosphorus level isn't raised to balance the ratio.

Diagnosis of Imbalance: Based on ration analysis. Evaluate the feeds and soils regularly, especially low-fertility soils with low pH (acidic). Rectal examinations by your vet may detect poor uterine tone after calving. Blood tests are occasionally used.

Treatment of Deficiency: Correct ration deficit (see sources). Feed high-quality alfalfa. Supplement vitamin D, especially if cows don't get much sunlight.

Calcium's Payback

Dave and I have always had a good rapport. Before starting a health program with me he never spent any money on vet services other than for milk fever or calving difficulties. He used to brag that his vet bill for the year was less than $100. He was so tight with his money, I used to accuse him of screwing his socks on in the morning to make him feel especially tight.

His milk production reflected his Scrooge-like ways. After more than a year I convinced him that if he used my nutritional services or a local nutritionist's services his milk production could jump by a third. Naturally he chose the nutritionist because he was cheaper. I didn't mind because the nutritionist was competent and the important thing was to improve profitability of the herd. I had to nag Dave pretty hard to get that far with the program.

Several months later I noticed that the cows' uteruses were not recovering after calving as quickly as they should have. On one of the visits for palpation, I asked Dave what the status on calcium was in the ration. He put me off with a vague answer, so on the next visit I pressed him on the calcium status again. He really jumped down my throat and told me that it was fine. He said he was tired of my being on his case about things that he had already taken care of. I knew I had stuck my foot in my mouth by asking. Not only was he having a bad morning, but he was tired of my telling him what he needed to do.

A couple of weeks later the nutritionist called me. He had investigated Dave's request to check the calcium and found that the calcium supplement wasn't being added to the grind-and-mix ration. Dave's helper who did the grinding apparently had decided that it probably wasn't needed. I felt vindicated. On my next visit to Dave's, he never mentioned getting the calcium levels adjusted. I didn't either but several months later Dave told me that I had really turned his herd around and he appreciated it. He needed to infuse only a few cows and his average days open had dropped 23 days.

Phosphorus (P)

Function: Energy utilization, proper reproduction. Critical for adequate trace mineral absorption. Involved in metabolic functions (acid-base balance, fatty acid transport, protein formation). Factor in appetite control.

Interactions: Close relationship with calcium (see Calcium).

Sources: Cereal grains, grass forages.

Supplements: Monosodium phosphate, diammonium phosphate, bone meal.

Requirements: Dry cows need Ca:P ratio of 2:1, so maximum of 100 g Ca/day means maximum 50 g P/day. Milking-cow rations usually have Ca:P ratio of about 1.5:1, but this is not critical. A 1300-lb cow needs 17 g/day for maintenance and 1.98 g/kg of daily milk production.

Signs of Deficiency: Inactive ovaries indicated by lack of heats. May have individual cases of ketosis. P deficiency is often associated with symptoms of other trace mineral deficiencies. May eat wood or dirt, drink urine.

Signs of Excess: High levels narrow the Ca:P ratio, so dry cows more susceptible to milk fever.

Causes of Deficiency: Cows fed high-quality alfalfa diets without P supplement. Tieup from other elements (diets containing iron at more that 600 ppm can make P unavailable).

Causes of Excess: Usually oversupplementation.

Diagnosis of Imbalance: Based on ration analysis. Depraved appetite and trace mineral deficiencies should raise suspicion. Blood test occasionally used.

Treatment: Correct ration (changing from alfalfa to grass hay may be enough). In extreme deficiency, vet can administer injectable P.

Magnesium (Mg)

Function: Vital for energy usage. Interacts with many enzymes needed for metabolism. Exerts powerful effect on nerve activity.

Interactions: Its action is interrelated to potassium and calcium. Mg deficiency can be inversely related to potassium levels.

Sources: Common element in feed. Widely distributed in plants and animals.

Supplement: Mg oxide.

Requirements: Dry cows: 0.2-0.4% Mg in ration.

Signs of Deficiency: Cow off feed. May appear malnourished or unthrifty, especially in late winter. Retarded growth, low production, ketosis. Cow can appear to have milk fever but really has tetany. This can occur at any point in lactation. May be uncoordinated, twitch, have convulsions. Cows on lush grass can get grass tetany, go down and die quickly. Others may have excessive udder edema, retained placenta.

Signs of Excess: Weakness, crampiness.

Diagnosis of Imbalance: Based on ration analysis, clinical signs or blood and urine tests.

Treatment: Supplement for deficiency. Give Mg oxide at 2 oz/cow/day for ketosis (not very palatable – try not to use topdress on parlor concentrate). Give intravenous Ca with added Mg for tetany. For grass tetany, intravenous Mg sulfate.

Potassium (K)

Function: Helps kidneys maintain water balance in body. Critical to muscle contraction (especially heart).

Interactions: Interacts with Ca, Mg and salt (sodium chloride). Has a direct influence on hormone that causes urine excretion. When hormone levels are high, kidneys excrete extra K, body retains water, which increases udder edema.

Sources: Essential element but considered non-critical because of wide distribution in nearly all feeds. Major source is alfalfa forages (alfalfa can have K level of 3%).

Requirements: 0.8-1.6% in ration.

Signs of Deficiency: Cows handle heat stress poorly when K levels are low. Also crampy when rising in freestalls. Tetany can occur in midlactation. Poor growth in young stock. Cows drink other cows' urine. Occasional tetany.

Signs of Excess: Newly fresh cows have udder edema. May have retained placenta, be unthrifty. May also lead to displaced abomasum.

Causes of Imbalance: Related to feed sources, and the ratio to Ca and Mg.

Diagnosis of Imbalance: Based on ration analysis, clinical signs, and blood test.

Treatment: Balance ratio with Ca and Mg. Give intravenous Ca with K for cows with clinical signs of deficiency. May need large oral doses in a cow down with a K deficiency. In cases of excess, it is hard to remove excessive K from diet if ration heavy on alfalfa haylage, especially if alfalfa is rich in K. Monitor K levels in soils. May take a few years to alter soil levels.

Sodium (Na) and Chloride (Cl)

Function: Salt helps maintain body's acid-base balance. Part of many metabolic functions.

Interactions: With K to maintain water and electrolyte balance.

Sources: Primarily supplement. Some NaCl in forages and water. No proof that cow definitely needs Cl supplement.

Requirements: Feed free choice. Milking cows eat what they need. Offer dry cows block salt only, as they may overindulge in late pregnancy. Young stock should have loose salt.

Signs of Deficiency: Depraved appetite. Aggressive attitude to get salt when empty mineral feeder is filled again. Depression, weight loss and rough haircoat. Reduced milk yield in long-term deficiency.

Signs of Excess: Refusal of feed high in salt. Cows drinking salt-contaminated well water have low production, look emaciated. Dry cows have udder edema.

Cause of Deficiency: No supplement offered.

Diagnosis of Deficiency: Based on observation.

Treatment: Offer salt.

Copper (Cu)

Function: Necessary for the manufacturing of red blood cells. Important to nervous system in growing, unborn calf. Vital for growing strong bones. Essential in development of hair pigmentation. Necessary for strong keratin in hoof. Critical for good immunity. Involved in converting beta-carotene to vitamin A.

Interactions: Reacts with molybdenum and sulfur in rumen; they can tie up Cu. Excessive Cu makes zinc unavailable. Cu can occur as dual deficiency with cobalt (see Cobalt).

Sources: Variable amounts available in forages.

Supplements: Cu sulfate, Cu oxide, chelated Cu.

Requirements: Must base recommendations with molybdenum and sulfur levels. 20 ppm in ration for milking cows. 15 ppm for dry cows and heifers (assuming molybdenum level of less than 1.5 ppm and sulfur level less than 0.25% in feed and 350 ppm in water). Note that there is fine line between deficiency and excess.

Signs of Deficiency: Anemia, lameness, heel cracks, foot infections, bleached haircoat, reddish cast to black and white Holstein (they have no red genes). Fragile bones in calves, susceptible to fractures. Newborn calf may be retarded, unable to nurse. Cow has low immunity against infections, heart attacks, diarrhea, show no heats, brittle haircoat. Cystic ovaries, infertility.

Signs of Excess: May have acute massive red blood cell destruction, liver disease. May die (sheep especially sensitive – never feed dairy feed to sheep).

Causes of Imbalance: Related to molybdenum and sulfur status.

Diagnosis of Imbalance: Never base diagnosis on just 1-2 tests. Most dependable test for deficiency is liver Cu analysis (I usually take sample on cull cows at slaughterhouse). Blood tests are less reliable. Ration analysis, including molybdenum in feed and sulfur in water. Hair analysis is reliable.

Treatment: Correct low levels in ration. With molybdenum and sulfur tieup, boost Cu levels above normal requirements. Give injectable Cu glycinate to correct deficiency, but will cause abscess or lump at injection site. May need to eliminate interacting elements (molybdenum, sulfur). Feed ammonium molybdate to correct toxicity.

Unraveling a Copper Mystery

Tim's cows had a rash of health problems. On my first visit to his place, I saw that many of his cows were thin and had draining sores. Some of the ugly festering areas were on the pinbones, some on the hipbones and some on hocks and knees. He's 6 ft 6 inches tall, and weighs 285 lb. He's only in his late thirties but he has lost most of the hair on his head. In the summer he worked without a shirt and is so big and hairy that he could pass for a gorilla.

Tim complained that his cows were lame and reluctant to walk across his gravel driveway. I picked up several of the cows' feet. I could see most had heel cracks and the hooves were soft and flaky. Tim was also having a tough time catching cows in heat. His cows weren't showing heat, even with heat detectors and a teaser cystic cow. The calving interval average was over 14 mo.

Alexia, Tim's wife, is a real knockout and a sun worshipper. She frequently comes to the barn in a halter top and short shorts. One time I had a male vet student with me when I visited Tim's farm. I could see when we arrived that the student was going to have trouble concentrating on our work. That was the case until Tim came around the corner without a shirt on. My vet student's eyes got big as saucers. Suddenly, he was extra polite: "Yes, sir" and "No, ma'am."

Solving Tim and Alexia's problems proved to be difficult and involved. First we started a herd health program. This was to establish a foundation of care so that we could be sure that the basics were being done. We found a number of problems. Nearly every cow had metritis after calving so it became routine to infuse every cow. Another problem was that production always started high but would never peak. Milk flow dropped steadily after calving. Cows freshened in good flesh but steadily got thinner. They had freestalls but usually preferred to be either in the manure storage or a muddy area adjacent to it.

Luckily, Tim was very cooperative. We vaccinated for all major diseases, such as IBR, BVD and Lepto. Later we vaccinated for Hemophilus. *Tim also talked to "experts" on* Chlamydia. *They agreed with Tim by telephone that his cattle had it. On their recommendation, Tim's herd had been vaccinated every other month with a* Chlamydia *vaccine. I also blood tested Tim's cows for most of the major diseases. But there were no clear-cut remedies to the problems plaguing this herd. Finally I decided that the ration needed a thorough evaluation.*

Tim had recently changed feed companies for the "nth" time. The previous feed supplier had balanced the ration with protein, energy, calcium and phosphorus. None of Tim's old rations had reported any trace mineral level. My tests revealed that the feed was adequate in all respects except for copper. Copper levels in the forages were low at 6 ppm. We then decided to blood test for blood copper levels and to sell a couple of cull cows on the rail. I went to the packing plant and got liver samples from the cows on the kill floor. My lab technician reported that the blood tests were marginal for copper and the liver samples were low also. Soon after that a corn silage analysis showed a copper content of only 1 ppm. At that point there was no doubt that Tim's cows had a copper deficiency.

Pieces of the puzzle were beginning to fit into place. Because of the copper deficiency, many of Tim's cows had bad feet. Tim's cows also had a reduced defense mechanism. Their impaired immunity resulted in continual health problems. I didn't see any calves in Tim's herd that were born too dumb to suck. But it wouldn't have surprised me if I had seen calves like that in the herd. Part of the reason Tim wasn't seeing his cows in heat is probably because without copper his cows were unable to convert beta-carotene to vitamin A. Tim had previously had a couple of cows fall over dead unexpectedly. At the time I couldn't explain it, but since low copper levels lead to scarring of the heart muscle and heart attacks, I suspect that it may also have accounted for the acute deaths.

Since molybdenum and sulfur can tie up copper, I thought it wise to check the levels of these elements in the feed and water. Neither was above normal. We raised the copper in the feed to 20 ppm for the total ration. The results of that move were encouraging. We no longer had to infuse most of the cows. Tim was able to see his cows in heat. New cases of cows with draining sores ceased to be a problem. The foot trimmer told me that on his regular visit the cows' feet were healthier. He didn't know what had been done but there were fewer heel cracks. Best of all, Tim's milk production went from 35 lb to 62 lb per cow per day during the next year and a half.

If you're wondering about my vet student, he now has his own practice and is quite successful. He recently told me he probably never would forget Tim's herd for more reasons than one.

Selenium (Se) and Vitamin E

Background: Nutrients function together and are found together in nature. Se, once thought to cause cancer, was illegal to add to rations. Later discovered as an essential element. Many areas of United States are Se deficient. Great Lakes region critically short, and 40 other states deficient in certain areas. FDA now allows Se supplementation of ration at 0.4 ppm (Fig 2).

Function: New research indicates Se is an anti-cancer agent. Involved with a number of oxidative processes. Especially involved with maintaining good immunity.

Interactions: Moldy feed, sulfur or zinc can make Se unavailable.

Sources: Feeds grown on high-Se soils don't provide enough. Sheep salt (higher in Se) isn't adequate source either. Purchase feed-grade sodium selenite.

Requirements: Dry cows need 3 mg Se/cow/day (large breeds) or 1-2 mg/cow/day (small breeds), 1000 IU vitamin E/cow/day. For milking cows, 5-6 mg/cow/day (large breeds), 3-4 mg/cow/day (small breeds), and 200-600 IU vitamin E/cow/day. These levels may not be sufficient when interactions with other minerals or moldy feeds tie up vitamin E or Se.

Signs of Deficiency: Embryo in pregnant cow dies (cow you thought pregnant comes back in heat in 40-60 days). In calves (white muscle disease), calf becomes unthrifty, may go down and can't get up. May develop ulcers, die of heart attack. Highly susceptible to diarrhea and

Figure 2. Areas in which forages and grains contain low, variable, adequate and toxic amounts of selenium. Selenium deficiency occurs in various areas of 45 states. (Courtesy of Diamond Crystal Salt)

pneumonia. On postmortem exam, muscles are pale, feel gritty. Cows often infertile, have poor conception rates, long calving intervals, higher susceptibility to infection. Herd may have high SCC. Dry cows freshen with retained placenta, later develop cystic ovaries. Some cases of fat cow syndrome related to this deficiency because liver can't provide Se and vitamin E for oxidative enzymes in metabolism. Fat cow or cow with retained placenta may then develop displaced abomasum. Low Se in unborn calf may cause defect in abomasal tissue, making it more susceptible to displacement as an adult.

Signs of Excess: "Blind staggers" when grazing plants on Se-rich soil accumulate the mineral. Accidental excess in ration leads to acute onset of respiratory weakness a few days after cow ingests Se. Goes down and can't get up, dies. Postmortem exam shows edema of the lungs. Some reports indicate that low chronic toxicity can cause scurfiness around the tail and head.

Diagnosis of Deficiency: Sometimes difficult to determine by observation alone. Blood serum tests or liver assay is dependable. Postmortem exam. Ration and soil analysis unreliable. Se in plants is often unavailable to cow.

Treatment: Effects of Se deficiency can be somewhat offset by feeding extra vitamin E, and reverse is also true; best to balance entire ration to meet requirements; injectable vitamin and Se given 2-3 wk before freshening help prevent calving-related infection but not suitable as long-term source.

Manganese (Mn)

Background: Levels in forages often border on deficiency. Poorly absorbed in digestive tract. Other elements can interfere with absorption.

Function: Critical for formation of normal cartilage. Also reproduction. Involved in several enzyme systems.

Interactions: When Ca, iron and K levels high, Mn is less available.

Sources: Present in most plants and forages.

Supplement: Mn sulfate.

Requirements: Most cattle need 60-100 ppm.

Signs of Deficiency: Inefficient breeding, poor conception rates, delayed ovulation, poor heats. Reduced body fat, slow blood clotting. Newborn calves may have defective cartilage in legs. Young cattle grow poorly, appear stunted.

Signs of Excess: Never observed in nature. Under research conditions, cattle grow poorly.

Diagnosis of Deficiency: Base on ration analysis. Hair analysis not reliable.

Treatment: Correct deficiency with dietary supplement.

Zinc (Zn)

Function: Maintains normal growth and health. Critical to healthy skin. Important in early pregnancy.

Interactions: High Zn in ration may tie up Se. High Cu or Ca can bind Zn and make it unavailable.

Sources: Found in low levels in roughages.

Supplements: Zn sulfate, Zn oxide, Zn methionine.

Requirements: Cattle need 80-120 ppm.

Signs of Deficiency: In severe cases, cows may be off feed. In marginal cases, cattle have chronic skin lesions, ringworm, rough haircoats and an unthrifty appearance. Reproductive failures, low conception rates. Swollen joints, lameness, abnormal hoof growth, other foot lesions. Cows salivate profusely while eating.

Signs of Excess: Cattle off feed, eating wood and dirt. Excess ties up Se, so cattle appear to have Se deficiency.

Diagnosis of Deficiency: Based on ration analysis, clinical signs and blood tests.

Treatment: Correct deficiency by adding supplement to ration. All forms equally available to cattle.

Cobalt (Co)

Function: Promotes growth of rumen microorganisms that convert Co to vitamin B_{12}. In high producers, only about 3% of B_{12} produced in rumen is absorbed.

Interactions: Excessive molybdenum causes Cu deficiency, and a Co deficiency occurs simultaneously. The conditions causing excessive molybdenum also cause Co deficiency. Soil pH above 6.7 (alkaline) caused forages to take up more molybdenum and less Co. Optimum dietary Co levels tend to enhance availability of Se.

Sources: Found in most plants (unless raised on soils with pH of 6.7 or higher).

Requirements: Most rations contain 0.1 ppm (NRC recommendation). I believe 0.5-1 ppm will lead to optimum production. Top producers often benefit from weekly injections of B_{12} to compensate for high requirement and low rumen absorption.

Signs of Deficiency: Related to vitamin B_{12} deficiency; anemia, low production, ketosis, unthriftiness, poor heats. In acute cases, cattle waste away.

An On-Farm Example

I think of Vernon's herd when I want to describe a case history of cobalt deficiency. Vernon is hard working, hard driving and stubborn whenever I try to tell him something. He's milking 80 cows, has an 18,000-lb average and is very particular about how his cows are cared for. Vernon always pays his bill but he always has a question about something I charged him for. When he doesn't want me to hear what he says to his wife and kids, he talks in German. Sometimes I think he's barking orders to them from the tone of his voice. Other times he is probably talking about me. Of course, I go out of my way to try to rattle his chain.

Earlier, we had recognized a copper deficiency in Vernon's herd and I had provided a way for him to correct the problem. He was metering copper sulfate into the ration and seemed to be getting a favorable response with this supplementation. But during one stretch in the summer, Vernon had about 20 fresh cows and it seemed that nearly every cow got ketosis about 2 weeks after freshening. We had so many cases of ketosis that it involved 15% of his total herd.

Generally it occurred in cows that were milking 80-90 lb of milk/day. They went off feed acutely, and sometimes showed nervous behavior. With intravenous glucose and other ketosis treatments they would straighten out for another 1-2 weeks, then lapse into ketosis again. A cow with ketosis might relapse 2-3 times before we finally got her healthy and milking. Of course, by the time we got them straightened out most of these cows would no longer peak in production. Overall, their production level was lower, too. Vernon was beside himself, and if Vernon was beside himself, I can assure you I was beside myself as well.

After much sleuthing, ultimately I looked at the cobalt levels in the ration because I know cobalt deficiency can be associated with a copper deficiency. Since there are no reliable diagnostic labs available in my area for cobalt analysis, we elected to do some cobalt supplementation. It wasn't until later that I learned of an inverse relationship between cobalt and molybdenum. I talked Vernon into adding cobalt carbonate to the ration at 0.5 ppm. I didn't know what else to do and Vernon was on my case. Within a week, the ketosis cases stopped. And, amazingly, the whole herd picked up in milk production. After a year the rolling herd average had come up 1500 lb. An added benefit was that Vernon's cows came into heat more reliably.

Vernon is one of those people who believe in doing the best they can in all aspects of an enterprise. He raises terrific alfalfa and puts a lot of lime on the fields to keep the soil pH up. In fact, the pH of his alfalfa fields was 7.2. The copper deficiency resulted from the high soil pH and increased molybdenum uptake in the alfalfa plants. At the same time, the high pH reduced the cobalt uptake in the forage.

Diagnosis of Deficiency: Base on ration analysis for Co and molybdenum. Few labs run forage analysis for Co. Urine test for malonic acid is an indirect measure of Co, not yet available in labs. Co deficiency may increase incidence of Johne's disease in herd..

Treatment: Correct deficiency by adding Co to ration. Acute cases (cows in ketosis) may need vitamin B_{12} injections. Co drenches and bullets for rumen also available but not necessary if diet balanced.

Iron (Fe)

Function: Works with numerous enzymes. Necessary for formation of hemoglobin (carries oxygen in red blood cells).

Interactions: Competes for absorption sites in intestines with Ca, P, Zn, Cu. Also has special relationship with Cu because a Cu-containing enzyme makes Fe available for hemoglobin. Low Cu causes poor Fe utilization, anemia.

Sources: Readily found in nearly all roughages, cereal grains. Cattle grazing on roughages, pasture get enough Fe (about 150 ppm). Milk is low in Fe.

Signs of Deficiency: Anemia, unthriftiness, increased respirations. Veal calves susceptible when on milk-only diets. Heifers can suffer if heavily infected with worms that suck blood in digestive tract. Cow with blood loss may then suffer Fe deficiency.

Signs of Excess: Extra Fe competing for absorption sites may cause deficiency of other elements (see Interactions).

Diagnosis of Deficiency: Blood test.

Treatment of Deficiency: Use Fe dextrans or gleptoferran injection. Occasionally need blood transfusion.

Iodine (I)

Function: Necessary to produce thyroid hormones.

Interactions: High Ca interferes with I absorption.

Sources: Readily available in iodized salt.

Requirements: Cattle need 0.6-0.8 ppm in ration.

Signs of Deficiency: Stunted growth, low milk production. Cows won't show heats. Poor conception rates. In advanced cases, may have goiter under chin (enlarged thyroid). In pregnant cows, calves born full term but hairless.

Signs of Excess: Reduced immunity against disease. Lameness, poor production, high SCC.

Cause of Imbalance: Low iodine levels in soil lead to deficiency, especially in Great Lakes states. Excess stems from offering several feed ingredients, each with added I.

Diagnosis of Imbalance: Based on signs and ration analysis. Blood test.

Treatment of Deficiency: Use iodized salt, supplement with an organic iodide called ethylene diamine dihydroiodide (EDDI).

Treatment of Excess: Remove excess from diet.

Sulfur (S)

Function: Needed for manufacture of amino acids from ammonia by rumen microorganisms.

Interactions: In presence of molybdenum, can bind Cu so it's unavailable. High S levels can also tie up Se. S can tie up other minerals when S level in water is 350 ppm or higher.

Sources: Commonly found in legume roughages, but low in corn silage.

Supplements: Available as a sulfate, such as Ca sulfate (gypsum).

Requirements: Cows need 0.22% in ration. Toxicity can occur if level exceeds 0.35%. Balance S in a 10:1 ratio with nitrogen sources.

Signs of Deficiency: Low production from poor protein utilization. Reduced feed intake. Growing heifers gain slowly.

Signs of Excess: Makes Cu, Se, Fe, Ca unavailable.

Causes of Imbalance: Low S in forages leads to deficiency. Excessive S supplementation or highly sulfated drinking water leads to toxicity.

Diagnosis of Imbalance: Base on ration and water analysis.

Treatment: Correct ration. In extreme cases may need to put S filter on the water source.

Vitamins

Vitamin A

Function: Critical for normal eyesight. Needed for bone development and maintenance of membranes of nostril, vagina, etc. Helps reproductive functions.

Interactions: Precursor is beta-carotene (found in most green forages). Enzymes convert it to vitamin A or body absorbs carotene intact and other tissues convert it to vitamin A. Enzyme that does conversion contains Cu, so deficiency in Cu suggests a deficiency also in vitamin A. Moldy feed can destroy vitamin A. Nitrates in feeds can tie it up.

Sources: One of the richest sources – polar bear liver (your trivia for the day). Vitamin A is usually purchased as a premix.

Requirements: Dry cows: at least 50,000 IU vitamin A/day. Milking cows: at least 100,000 IU/day, in some cases 2-4 times that amount.

Signs of Deficiency: Depressed feed intake, slow gains. Night blindness, excessive tearing of eyes, occasional diarrhea. Poor conception rates, poor corpus luteum development on ovaries, reduced progesterone production. Calves born of deficient cows may be weak, may have birth defects. Possible embryonic death.

Signs of Excess: Rare. Staggering, convulsions, death.

Cause of Deficiency: Beta-carotene and vitamin A dissipate from green forages when they're ensiled or stored in a mow. Can happen in 6 months. Deficiency will probably show up in February or March. Also, hay loses beta-carotene when it's rained on. Excess: Accidental over-supplementation.

Diagnosis of Deficiency: Based on ration calculations, signs, blood test (samples must be frozen within an hour after collection). Deficiency signs in a well-fed herd may indicate forages are high in nitrates.

Treatment of Imbalance: Supplement ration. Large doses of vitamin A can override tie-up by nitrates. Injections of vitamin A in oil for immediate response are inexpensive, last 6 months. You may prefer twice-a-year injections to supplementing rations.

Beta-Carotene

Function: See Vitamin A. Some research indicates supplements improve conception rates and milk production.

Interactions: See Vitamin A.

Sources: Green forages.

Requirements: German researchers believe high-producing dairy cows may need supplements of as much as 1000 mg/head/day in winter. Other researchers' work has not confirmed this.

Signs of Deficiency: Same as for vitamin A.

Cause of Imbalance: Poor feed quality, especially in late winter.

Diagnosis: Blood tests seldom used.

Treatment: A few experts recommend supplementing 300-1000 mg/head/day in winter months but many others question this.

Vitamin D

Background: Actually a hormone. Ultraviolet light from sun stimulates skin to synthesize vitamin D.

Function: Stimulates sites in intestine to absorb Ca. Influences P absorption.

Sources: Sunlight stimulation. Premix supplements.

Requirements: For cattle housed inside, without access to sun, add 20,000-30,000 IU/head/day to ration. Never exceed 40,000 IU/head/day.

Signs of Deficiency: Reflected as Ca-related illnesses. In severe cases, bone defects in mature animals. Bones demineralize, fractures result. In young animals, bones grow crooked.

Signs of Excess: Ca deposits in kidneys, arteries, heart. With severe toxicity, death.

Causes of Imbalance: Deficiency: Difficult. Rule out other possible causes, then observe changes indicating Ca- and P-related illnesses.

Treatment: Correct ration. Allow cattle access to sunlight.

20

Effective Feeding and Watering

You may think you already know enough about feeding cattle. This may be true, but perhaps you can improve your setup and ration so that you can hike profits by increasing efficiency.

Often I hear clients say they're feeding cows one way or another "because this is the way that we've always done it." I'm challenging you to evaluate that, and I hope you'll see that you won't necessarily have to overhaul the way you do things to improve your herd's performance. There are some really basic, simple things you can change to get positive effects.

Feeding Individual Cows

Parlor Feeding

Parlor feeding is advantageous because you can correlate feed with production and your cows are quieter while you're milking them. One disadvantage is that cows don't have enough time to eat the amount of feed they need. They tend to gorge themselves but don't eat enough for highest productivity.

When cows gorge themselves in the parlor, rumen acidosis tends to be a problem. The condition shows up when a cow eats a large quantity of grain in a short time and doesn't get enough fiber. Using feeders in the parlor also costs you in equipment and maintenance. The humid atmosphere makes the metal parts rust out quickly, and dairymen say they must repair the equipment frequently. If you are redoing a parlor or installing a new one, it's considerably cheaper to leave out the feeding equipment.

Feed separation is always a problem where augers deliver feed to the parlor. This is because the fine material is left behind. And when the finely ground feed does make it into the parlor, cows don't seem to want to eat it out of the trough. Feed in parlors also makes lots of dust. If your milking equipment filters are also in the parlor, they may get plugged with dust.

Magnetic Feeders

Another individual feeding system is a magnetic feeder. With this system, a cow wears a cord or chain with a magnet on it around her neck. She can put her head in the feeder and as long as the magnet makes contact with the electrical plate or the chain makes contact with the magnet in the plate, feed continues to dribble to the cow. It's an effective way for you to deliver extra feed to your high producers, and it's relatively inexpensive. Dairymen say the feeders are pretty easy to install.

Some dairymen use a special feed blend in their magnetic feeders. I recommend one-third corn, one-third soybean hulls and one-third dried distillers' grain. Cows tend not to overeat with this blend. This mixture, though, may not flow well through some feed systems.

There are some disadvantages. First, you must change magnets regularly. Also, you must continually keep track of which cows are high producers and which ones have slacked off. When you leave lower-producing cows on the feeder, they get fat because they eat more grain than their production warrants. This fattens your feed bill, too! Another disadvantage of magnetic feeders is that they must be operational 100% of the time, or the cows become conditioned to an absence of feed and stop using them. Every time they put their heads in that feeder, they should get feed.

Another disadvantage is that cows tend to overeat at the magnetic feeder. When they get too much grain, they're prone to acidosis and laminitis. You'll also notice a drop in your butterfat test.

Computerized Feeders

Computerized feeders operate like magnetic feeders in that each cow wears a cord with an identification code around her neck. When a cow puts her head in the feeder, the scanner on the front of the computerized feeder interprets the code on her neck and identifies the cow. You program the computer to feed each cow a specific amount in a 24-hour period. The computer activates the feeder and delivers the cow's quota. In most systems, you program how many portions the cow receives in a day. For example, if you say she ought to eat 5 times a day, she'll get one-fifth of her quota each of 5 times she goes to the feeder. After that, she won't get any more feed (Fig 1).

There are a number of advantages to computerized feeders. Problems with rumen acidosis are greatly reduced by not allowing cows to gorge on a large quantity of concentrate at once. If you've programmed the computer correctly, it's impossible for the cow to overeat. A cow can be fed in relation to her production and condition. It's an excellent means

Figure 1. Computerized feeders are a great improvement over the older magnetic feeders because they precisely control the amount of feed delivered, and provide valuable information on feed consumption by each cow. Each cow wears an electronic identification code in a neck tag. When the coded neckband touches the front of the computer feeder, an electronic eye reads the code and delivers the appropriate amount of feed for that cow and records her consumption. (Courtesy of Germania Dairy Automation)

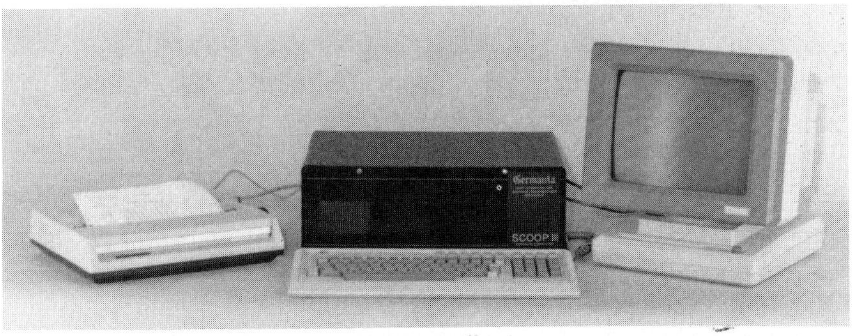

of measuring a cow's daily intake because the computer will print out how much feed each cow eats at each feeding. Consequently, this is an excellent means for detecting when a cow is off feed from illness or heat.

One of the biggest advantages of computerized feeders is fast payback. Most computerized feeders give a 12- to 15-month payback if you install one as a replacement to a magnetic or parlor feeder. I can't think of anything else in dairying that will give as quick a payback as a computerized feeder.

The drawbacks to these feeders include a high initial cash outlay and continued maintenance expenses. You must monitor them daily. The computer requires regular calibration. Lightning and other disturbances can throw the calibration off so the computer is delivering a different quantity of feed than intended. I suggest you recalibrate with every new batch of feed. There's nothing magical about computers, though. Unless they are used properly, they will not help.

Compared to magnetic feeders, computerized feeders lead to few digestive problems. Rumen acidosis can still occur, however, as evidenced in butterfat suppression commonly seen in cows on a computerized feeder.

Computer Error

While Ron and Nancy never had a fancy herd average, they always worked hard and tried to do a sound job of the basics in managing their dairy herd. Several years ago they decided to purchase a computerized feeder. They had it installed and were excited about what their computer could do.

Indeed, their cows picked up about 6 lb in production per day. The couple noted that their feed costs dropped by a fourth, as compared to when they used their old magnetic feeder. They were excited.

But after several months Ron and Nancy noticed their cows were getting thin. They had a computer station for their dry cows too, but they were looking thin also. They asked me to look over their ration after their nutritionist did. When I did, I told them I thought the nutritionist had done a good job. I could see nothing wrong with the ration. Finally I told them that they couldn't rely solely on what the computer dictated they put in the feed. They needed to develop a critical eye for their cows' condition.

My eye told me that their cows were getting thin despite their getting the proper amount of feed, so I advised them to increase the amount of grain each cow was delivered from the computer. It wasn't until we doubled the amount of grain each cow could get from the computer that we felt we'd done enough for the cows.

At that point milk production increased and the dry cows began to have better flesh. I still couldn't understand why these cows required twice as much grain as any other herd of comparable production. Later the nutritionist and I discussed Ron and Nancy's herd and the problems they were having maintaining production and flesh on their cows. The nutritionist told me that he had been there 2 days previously and hit upon the idea of double checking what the computer was doing. He recalibrated the computer and discovered that for the past 6 months the computer had only been delivering half as much feed as what it was programmed for. When I had advised Ron and Nancy to double the ration through the computer, I had only raised the ration back to the level that was required for optimum production!

I was astounded at what the nutritionist told me. On my next herd visit to Ron and Nancy's farm, I asked them about the computer and the nutritionist's findings. They were puzzled about having to double the ration, too, but finally recalled a terrible thunderstorm about 7 months earlier when lightning had struck close to the barn. Undoubtedly the lightning had knocked the computer out of kilter.

Tie Stall and Stanchion Feeding

Feeding in tie stalls or stanchions is probably the ultimate in individual care for each cow. Many purebred breeders like this setup because it allows them to keep excellent individual records. The cow gets individual care and you can monitor her feed intake closely. Tie stalls also eliminate the feeding problems common to other setups because cows don't have to compete for feed.

A disadvantage to this kind of system is that it requires a lot of time and labor. If you manage a commercial herd, tiestalls are probably not the best choice. However, if your herd is purebred, and records of high production are a part of selling breeding stock and adding to farm profits, the extra labor and time you must invest in tie stalls are well worth it. The only other disadvantage is that it's harder to detect cows in heat unless you turn them out for several hours every day.

Feeding Groups of Cows

Bunk Feeding

One way of feeding a group of cows is simply to have one bunk feeder for the group. On the positive side, it requires considerably less labor to feed a number of cows at one time. You have a harder time feeding minerals, however. It is nearly impossible to customize your feeding program for the production level. This usually means feeding minerals free choice (see Chapter 19).

I don't recommend free-choice feeding because cows have a tendency to overindulge in palatable minerals. Conversely, they won't eat enough of a mineral supplement if it is not very palatable. You end up feeding

Figure 2. This portable mixer is for making a total-mix ration. All of the corn silage, grain, minerals and byproduct feeds are blended together for a total mix. Note the scales on the front for precise addition of each ingredient.

all cows based on the average for the herd, rather than tailoring mineral levels to the top or bottom groups.

Total-Mix Ration

With a total-mix ration (TMR), you use a mixer to combine all the grain concentrate, minerals and high-moisture forages into one blended feed. This works well to eliminate rumen acidosis. The cow eats fiber and concentrate together. Finicky cows don't have any choice about what to eat and what to avoid. I'm sure you have cows that like haylage better than corn silage, or hay better than haylage. With a TMR, they must eat what's put in front of them (Fig 2).

Most feed blenders have scale systems on them, so you can keep an eye on how much feed a group of cows eats. It's also easy to feed for production level. You can split the herd into 2 or more groups and blend different batches of feed for each.

Keep some basic feed ingredients, such as corn silage or haylage, as the foundation for your ration, and then you can change other feed ingredients without the cows' detecting it. That way, cows won't go off feed or drop production every time you take advantage of a good buy on another ingredient. In today's marketplace, there are often excellent alternatives in byproduct feeds to boost production and cut feed costs.

Blenders are a very efficient method for using byproduct feeds, such as cottonseed, brewers' grain and distillers' grain. Some of these feeds, particularly cottonseed, may not be palatable if fed alone. But by mixing them with haylage or corn silage, you mask the taste and deliver the benefits of byproducts. TMR also is an excellent means of feeding vitamins and minerals.

If you're still relying on an individual feeding system, a mixer could be one of the best additions to your farm.

There are some disadvantages to TMR. The thing that will probably get your attention first is the big price tag for a mixer. You must also monitor and weigh all the ingredients that go into a mix. You'll probably spend more time feeding, too, because you have to feed the herd in groups. For these reasons, a mixer is rather impractical for a small herd. A mixer may not be appropriate for you if you're feeding longstem dry hay, either. You should double check that your mixer can handle hay. If it won't, then feed hay separately or chop the hay before it goes into the mixer.

Combination Feeding Systems

Some dairymen are happy with a "marriage" of individual and group feeding systems. I give high grades to a TMR if you have a large herd divided into groups of cows with similar feeding requirements. If this isn't applicable, my second choice is a TMR and a computerized feeder. This is nearly as effective for feeding dairy cattle in groups. Some of my clients can't justify the expense of a mixer and a computerized feeder, though, so they put in a computerized feeder to work with bunk feeding. This works pretty well, but rumen acidosis can be a problem because the grain isn't mixed with the forage.

The dairymen I know who use bunk feeders and magnetic feeders together complain that their cows have stomach disorders. Laminitis (founder) is common, and feed costs are high. This kind of combination makes it hard to reach your optimum profit potential.

With bunk and parlor feeding, you give up profits because you can't control the feed and minerals effectively at either site.

The Importance of High-Quality Water

Take some time to evaluate the quality of water you provide for your herd. Many dairies use ponds, streams and canals to water cattle. This is undesirable for several reasons. These sources are easily contaminated by wild animals that may carry leptospirosis or other infectious diseases that can make the cattle sick. These water sources also may be contaminated with nitrates, nitrites and coliforms. It is common for a herd to have nonspecific health problems when its water is contaminated in this way. A stream may also carry contaminants from neighboring dairies. Ponds are reservoirs of disease, and stagnant water is never good for a dairy cow.

If you don't have adequate well water supplies, you have no choice but to use ponds and streams to water your cattle, or to haul water in from local municipalities. This is a poor choice. Talk to your vet about getting your water checked for quality and addressing water contamination problems, whether they involve nitrates or coliforms. Have the water tested by an independent laboratory. If necessary, install a chlorinator on the water line if the water can be pumped to a watering tank. The chlorine then dissipates from the water as it is pumped into the waterer.

In some areas dairymen have elected to go to commercial water sources. These cooperative water companies have water lines available up and down the road for each client. These water sources tend to be more expensive; however, they're certainly a much better choice than having continual disease problems and low production from contaminated water.

Artesian well water is another source for some dairies. There are only 2 concerns that I want to mention with this source. You may have polluted ground water, and you'll tend to have high mineral levels in the water. By all means, have the well tested so you know what you're dealing with.

Another water source is a drilled well. Drilled wells are expensive if you have to go very deep. At my own home, the well is only 15 ft deep, but it's drilled into gravel, which acts as a filter for any water coming in. The main concern in this case is contamination of surface water flowing into wells.

Deep wells also are notorious for high sulfate levels. Wells 500 ft deep or deeper may have sulfate levels of at least 1000 ppm. Sometimes drilled wells develop a hole in the well casing. This allows surface runoff of coliforms and other contamination into the well and leads to a contaminated supply similar to that from ponds and streams. The only solution is to drill a new well. You should also be aware that you can buy well casing made from PVC that won't rust. A temporary solution to contaminated well water is to weekly add a gallon of Clorox to the well and not pump from the well for 12 hours. Then you pump until the water quality looks okay. By doing this weekly, you reduce the contamination levels until you're ready with a more permanent solution.

Watering Systems

There are a number of different watering systems on the market. Each one works in a different situation.

Water Tanks

Water tanks are one means of supplying water. Make sure you provide enough capacity so that every cow can get her fill. A group of 35 cows needs at least one 300-gallon tank.

Automatic Waterers

If you've installed automatic waterers, make sure there's one for every 25 cows. Keep in mind that manure can contaminate these water sources, so keep them clean.

You Expect Us to Drink That?

One of the farms that I visit regularly is managed by Lowell, a herdsman for an absentee owner. Lowell used to see that the cows got fed properly, but he never was one for cleaning up the place much. When Lowell milked at this farm, manure would get stacked up everywhere. One time when I was in Lowell's barn lot I happened to look in his water tank. He had a 200-gallon tank for 70 cows, which was inadequate as far as I was concerned. It appeared shallow, with only 6 inches of water in it. Upon closer inspection, I realized that the tank had about 18 inches of dirt, manure and filth in the bottom of it, with the water sitting on top.

I plunged my arm down into this muck and discovered I'd guessed correctly. This tank hadn't been cleaned in a long time. I pointed this out to Lowell. He said, "Yeh, Doc, I just haven't got to that lately, but I'm going to get it done in the next day or 2."

A month later I was back and happened to be walking out through the barn lot again. I made a special point of looking at the tank. Lowell hadn't done a thing with it. I mentioned it to Lowell again. He said, "I've been busy with corn harvest, Doc. I'm going to get it done here shortly." I didn't respond.

The next month, I knew Lowell probably hadn't touched that tank, but I was determined that if nothing else happened on that farm, the tank was going to get cleaned. I made an excuse to go back out past that watering tank. I glanced into it and saw that, true to form, Lowell hadn't touched it. Lowell saw me looking at it, and I kept my mouth shut. I just turned around and walked back to the truck. Lowell asked me where I was going and I told him I needed some material for a talk that I would be giving to a group of dairymen. I went to the truck and got my camera and proceeded to take pictures of that tank from all angles. Lowell never said a word.

I wasn't sure whether I had gotten Lowell interested in cleaning the tank or not. But the following month when I was there on my regular visit I made an excuse to go out past the tank again. There it was—bright, shiny and clean with fresh water clear to the bottom of the tank. Too bad I hadn't thought up that trick several months before.

Stray Voltage and Water Intake

Stray voltage results from faulty equipment and can cause a number of problems in cattle. Research from universities has indicated that stray voltage does not reduce milk production and does not cause significant health problems. Most of this research has been done on 2-3 cows that were shocked intermittently for 6, 8 or 10 weeks.

In the field, however, I haven't observed this to be the case. There are documented cases showing that cows *can* have lower production from stray voltage. Stray voltage can come onto your farm via the neutral line from the power company. Ungrounded equipment on your dairy also can carry the stray voltage to the cows. Short circuits in water heaters frequently cause this problem. Sometimes the stray voltage may come from a neighboring farm. In one more notable case that I've seen, a farm about a mile down the road had a short in an automatic water fountain heater, and stray voltage traveled down the power company's wire to his neighbors' place a mile away and stressed the cows.

Give Me a Buzz

One of the classic cases of stray voltage I have seen was on Doug's farm. He called me because he was puzzled by cows that only kicked their milkers off at certain times. He asked if I'd come out and take a look.

I went to the farm and watched Doug milk. At first the cows came in okay and stood for milking. Then suddenly they all started prancing around, kicking the milkers off. I couldn't figure out why they felt discomfort all at once. Then, faintly in the background I heard a telephone ringing in Doug's office. It turned out that the telephone company had recently grounded Doug's new line to the hot water line that ran out into the parlor. When the phone rang, the current grounded on the water pipe and made the cows kick and jump around.

Signs of Stray Voltage

The signs of stray voltage are quite variable. Some cows may tolerate stray voltage and not show signs. Other cows may be nervous and excitable, and have poor milk letdown. Some cows have high somatic cell counts. If stray voltage is high, cows may refuse to eat in the parlor. Milk production may be depressed. Production can also be depressed from the fact that cows aren't letting their milk down. If the stray voltage is at the water source, cows refuse to drink. This also causes reduced production.

The signs are so variable that the only way you can be sure about stray voltage is to call in an electrician to check things out. Sometimes your veterinarian can help with the diagnosis, too.

One way to identify stray voltage is to drive a ground stake away from the milking parlor. Run a wire with a volt meter to the parlor, attaching the positive lead to various pieces of equipment in the parlor and measuring any voltage. Cows generally can stand up to 1 volt without showing outward signs of distress. Cows that are getting shocked with 4-5 volts are likely to be nervous and excitable.

Stray voltage is usually worse on rainy days. If the day is bright, sunny and dry, an electrician is not as likely to find the problem as when the weather has high humidity and is dreary and rainy. I know all the experts won't agree with my comments on stray voltage, but my observations are based on a number of herds that have suffered stray voltage.

Basically, there's only one remedy for stray voltage: find the source and eliminate it. Also, ground every piece of feed and parlor equipment. Then have the power company put an isolation transformer on the incoming power line that will stop any stray voltage from coming to your farm from the neighbor's. An equipotential grid mat buried in the parlor floor may also be needed.

21

The Future of Dairying

As we approach the year 2000, dairying is evolving into a new type of business. Dairymen who stay abreast of and use new technologies will increase profits and have a bright future in dairying. Others will have difficulty making a profit.

Change will be the key word for the new era in dairying. We can expect 100 lb of increased production per cow per year because of genetic improvements. Improved feeding practices, in conjunction with sharpening our management skills, will likely add another 150 lb per cow per year. Such innovations as bovine growth hormone used in combination with improved feeding and management practices will likely make the average cow of the year 2000 produce 20,000 lb of milk in a lactation.

But this kind of production increase will mean stiff competition for those of us who value the small family farm. Larger, well-managed dairy farms are often able to adopt new technology much more quickly. With this advantage, the larger farms may be in a much more competitive position to turn a profit as we enter the 21st century.

Experts estimate that by the year 2000, the industry will need 30% fewer cows and 51% fewer dairy farmers to meet production requirements. I want to emphasize that if you are serious about dairying, you will have to really hustle to be among the 49% of the dairymen still needed.

Here It Comes, Ready or Not

Get ready for the 21st century. Plan to be a winner. Be ready for the changing technology as it arrives. The revolution in agriculture is going to happen whether you are ready or not. It's a waste of valuable time to sit and castigate the bureaucrats for the problems you had in the 1980s. If you're going to be in dairying in the 21st century, you don't have time to waste. The year 2000 is a reality. If you're not willing to lay the groundwork for a thriving, profitable business of the future, and *milk your cows for all they're worth,* then move quickly to some other career option.

Here's a look into the future. Don't let this look into the future rattle you. Such innovations as artificial intelligence, biotelemetry devices and robot milkers aren't going to replace you, Mr. Dairyman. Rather, they will free you up from the tedious, time-consuming jobs, provide you with a wealth of useful information, and enable you to spend more time analyzing information and making more profitable decisions.

Advances in Genetics

The genetic makeup of dairy cows will continue to improve as geneticists develop improved methods of mating cows. Eventually most of the major genetic advances will come when only sires are used for parents. The breeding gains will come much faster than with some of the current techniques, such as cloning, because we already have so much statistical data on bulls. Their prepotency measurements are also much more accurate than current cow data. The disadvantage of cloning as compared to "mating bulls to bulls" is that the cloned calf can only be as good as the best of each individual parent. But by using 2 sires to get a calf, geneticists can combine the best of the most elite bulls. The technology already exists for this feat because of the recent advances in staining and identifying cattle pronuclei that contain the genetic material for the new calf.

Following recovery of a genetically superior embryo in an embryo transfer procedure, experts will perform microsurgery on the embryo. First they will remove the female pronucleus, leaving only the male pronucleus. Then a pronucleus from another genetically superior sire's embryo will be inserted into the embryo that contains the first male pronucleus. Next, the experts implant this embryo containing the 2 pronuclei from different sires into a recipient cow. The individual resulting from this pregnancy will in essence be a "bull bred to another bull." The newborn calf will have the superior genetics of both sires. Genetic gains from mating bulls to bulls will result in genetic improvement about 1 1/2-2 times faster than when bulls are mated to cows (Fig 1).

In addition, geneticists in the future will be able to map statistical predictions and plan the best matings between individuals who are far superior to the average individual of that particular blood line. For example, when a top cow from a special blood line has a heifer calf that's designated for a special genetic program, surgeons would remove the calf's ovaries at birth and harvest the 300,000 or so eggs present. They would fertilize them in the laboratory and put them into open recipient cows that lack the calf's superior genetics (Fig 2).

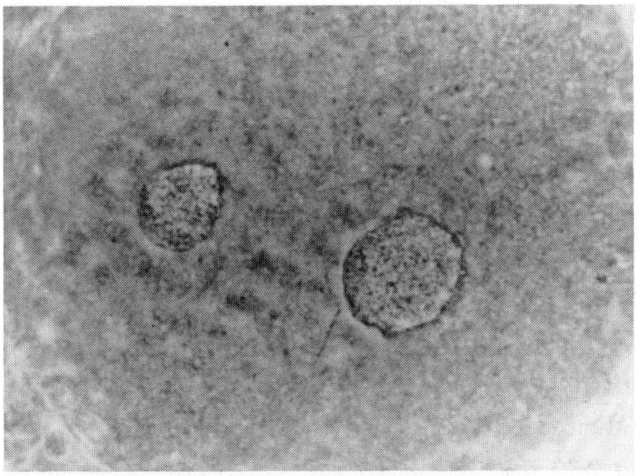

Figure 1. The pronucleus of a fertilized egg contains the genetic code. The pronucleus can be transferred to another embryo to influence the development of that embryo. In this way, scientists can use the genes of only one parent to influence the developing offspring. In the future, the surgical transfer of pronuclei may be routinely used to improve the quality of herds. (Courtesy of Dr. Neal Ferst, University of Wisconsin)

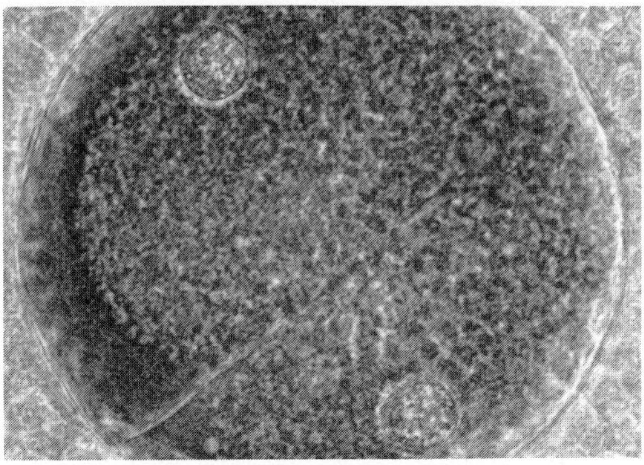

Figure 2. *In-vitro* fertilization, also known as test tube fertilization, eventually will be used to improve the genetics in new generations of bulls. (Courtesy of Dr. Neal Ferst, University of Wisconsin)

Breeding

A breeding program for your 21st-century herd may run like this. You would synchronize the heats in your cows so you could do all of the reproductive work on them once a month. You inseminate the top 1% of your cows with sperm containing the male Y chromosome. This would be the foundation for further development of new sires and bloodlines. Then when you wanted heifers for replacements, you'd breed cows with sperm with the X chromosome. By selecting the X chromosome, you would be assured of getting a heifer. Then you'd breed the bottom half of your herd with the Y chromosome and raise the calves as dairy beef. You may also opt to purchase frozen embryos to put in some of your cows.

Semen you purchase in the 21st century will be encapsulated for improved fertility. The encapsulation will have sustained release of the sperm to better coincide with the time of ovulation. No more worrying about timing of breeding. The cow is just inseminated when she's synchronized to come into heat.

Your vet will no longer be doing much pregnancy checking on your cows. Soon you'll be able to buy a new test kit that uses milk to diagnose pregnancy. This test will give a positive indication of pregnancy by detecting a substance associated with pregnancy called protein B. Researchers in Idaho are very close to working out the details.

In the 21st century you'll still need your vet to diagnose and resolve problems with cows that are infertile. However, this will be only a small part of his or her responsibility. You vet's time will be devoted primarily to production medicine, that is, to helping you improve the health and performance of your cows by updating management practices. Every management and health recommendation will have a price tag on it. In other words, before you accept and implement management advice from your vet, you'll assess it thoroughly for economic return and impact on the herd.

Computers on the Farm.

The 21st century will be called the "Age of Information." This should be no surprise to you. We are already inundated with digital watches with built-in calculators, computer games, and computer-generated billing and bank statements. It is no longer a question of whether we need a computer for our business, but how and when we are going to put one to work. I've heard it said that if the automobile industry had made the technological leaps that the computer industry has made during the past 30 years, a Rolls Royce would cost $2.50 and get 2,000,000 miles per gallon!

Your business' survival and prosperity will be determined by how quickly you can obtain and use new information. For this reason, one of your long-term goals should be to buy a computer terminal that's tied into a mainframe with readily accessible new data.

Your cows will most likely have electronic implants to provide identity for information collection. Every time she comes into the parlor a short-range, pulsed radar will record her identity. Her milk weight will be recorded, and she will have her temperature checked and SCC

Figure 3. The ZARTemp device contains a radio transmitter that transmits the cow's body temperature to a computer. It is placed in the vagina using a tube speculum similar to a balling gun. It floats freely in the vagina. The cow could be bred and maintain a pregnancy while wearing this device. It can also be used during late pregnancy to serve as an alarm for impending freshening. (Courtesy of Dr. David Zartman, Ohio State University, and Biomedical Electronics)

evaluated. You computer will report which cows are in heat while they are in the parlor (Fig 3).

Every time the cow goes to the computerized feeder, feed consumption information is collected. Automatic weighing systems are being developed that will automatically factor her weight into calculations of dry-matter intake and ration formulations.

Your computer system will establish a database on a diverse range of management factors. It will pull these various factors into an evaluation report that will greatly improve your ability to make management decisions based on current economic conditions. Artificial intelligence will match the facts to possible symptoms or conditions and use management rules to give you a probable course of action. Examples might include when to breed a cow, who to mate her to, and when to cull her.

Milking Equipment

Researchers are trying to develop a robot that will do the milking. Engineers have nearly solved the problem of how to get the milker attached. You'd have a new reason to cull a cow with only 3 teats because the robot probably won't be able to deal with the variation (Figs 4, 5).

Figure 4. Someday you may rely on a robot milker like this one. This system allows a cow to enter the parlor whenever she has the urge to be milked. The milker equipment starts, delivers feed to the cow, preps the cow and attaches the milker unit. European workers have found that most cows liked to be milked 4 times a day. Sensors in the system detect a high somatic cell count or fever. After milking, gates shunt a sick cow into a separate pen for examination by the herdsman. Robotic milking systems may be used on 10,000 farms by the 1990s. (Courtesy of W. Rossing, Instituut voor Mechanisatie, Arbeid en Gebouwen, Wageningen, The Netherlands)

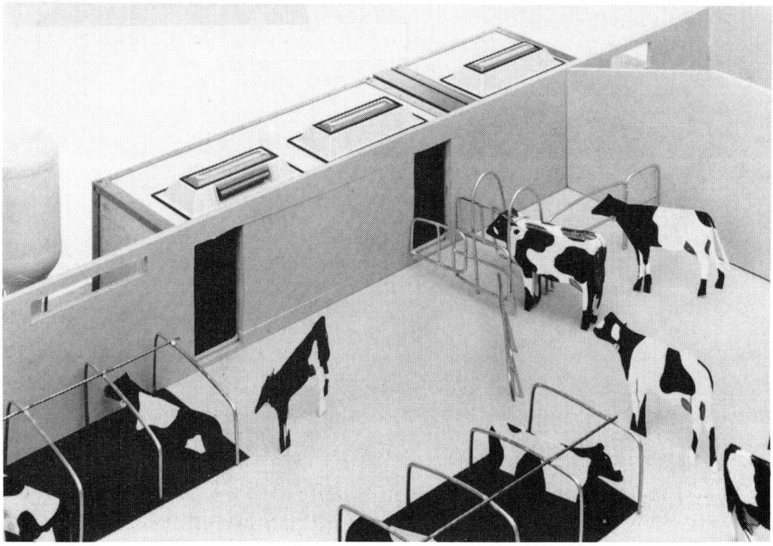

Figure 5. This is an actual working prototype robotic milker. (Courtesy of W. Rossing, Instituut voor Mechanisatie, Arbeid en Gebouwen, Wageningen, The Netherlands)

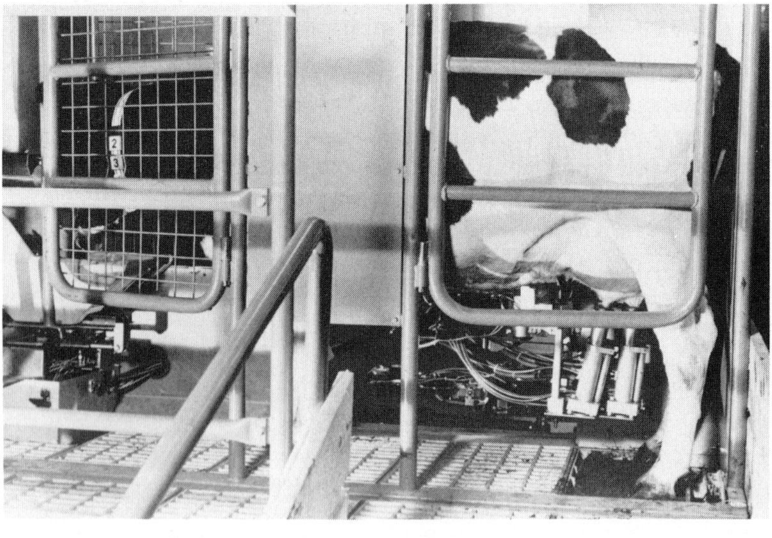

Another vision in milking equipment is hydraulic milking. This method of milk removal depends on keeping the teat end in contact with liquid milk rather than vacuum. This concept saves the teat end from the stresses of vacuum. Consequently, milker-related mastitis could be a thing of the past.

Mother Nature's Most Perfect Food

In recent years the dairy industry has done an excellent job promoting the health benefits of milk. You are familiar with the copyrighted symbol of "real dairy products."

The dairy industry will have problems maintaining credibility for producing a pure product if many of us do not change our milk production methods. The public is becoming very sensitive to chemical contamination in the food chain.

Antibiotic residues will continue to be a major concern and require all of us to be more innovative in dairy management. Withdrawal times for residues will become longer as more residue information is supplied to us. In addition, better methods of establishing withdrawal times on each individual cow will be available because of more cowside test kits.

I see no need to legislate a ban on bovine growth hormone for economic reasons. It isn't likely to be a part of the 21st century. Regardless of how many researchers have pronounced it safe, the public perception is it will contaminate milk. Dairymen have developed a wonderful credibility with the American public that milk is pure and safe. They cannot afford to taint that image.

Dairy co-ops will eventually choose to purchase milk that they can certify as being free of contamination. The public will likely pay a premium for a certified pure product. This will probably be the death knell of bovine growth hormone as it is projected now.

Other methods of naturally stimulating a cow's endocrine system will be developed with less fanfare than the choices now available.

Epilogue

Having read this book, you may feel overwhelmed by the number of recommendations you could implement in your operation. But you need not feel inadequate to the task. I'm betting that you are the kind of dairyman that can stand back and be analytical, then "go for it." Make a list of management areas needing improvement that won't cost much to change.

You may benefit by discussing these management changes with your vet. Not only will you gain his support, but also he may be able to suggest a faster method of making the change. Then, you should number each item in order of which will give you from the greatest return from the least effort.

Then select only 1-2 areas that will obviously give the best return for your 100% management effort. After you have this management problem area mastered, go for the next item on your list. Just remember, it isn't how much money you have to spend on improvements, but how you best use the resources you have.

Good luck and Godspeed on your journey toward boosting your dairy's profits.

Don Sanders, DVM

Index

G

H

I

J

K

L

M

N